VOLUME ONE HUNDRED AND THIRTEEN

ADVANCES IN
APPLIED MICROBIOLOGY

VOLUME ONE HUNDRED AND THIRTEEN

ADVANCES IN
APPLIED MICROBIOLOGY

Edited by

GEOFFREY MICHAEL GADD
Dundee, Scotland, United Kingdom

SIMA SARIASLANI
Wilmington, Delaware, United States

ACADEMIC PRESS
An imprint of Elsevier

ELSEVIER

Academic Press is an imprint of Elsevier
50 Hampshire Street, 5th Floor, Cambridge, MA 02139, United States
525 B Street, Suite 1650, San Diego, CA 92101, United States
The Boulevard, Langford Lane, Kidlington, Oxford OX5 1GB, United Kingdom
125 London Wall, London, EC2Y 5AS, United Kingdom

First edition 2020

Copyright © 2020 Elsevier Inc. All rights reserved.

No part of this publication may be reproduced or transmitted in any form or by any means, electronic or mechanical, including photocopying, recording, or any information storage and retrieval system, without permission in writing from the publisher. Details on how to seek permission, further information about the Publisher's permissions policies and our arrangements with organizations such as the Copyright Clearance Center and the Copyright Licensing Agency, can be found at our website: www.elsevier.com/permissions.

This book and the individual contributions contained in it are protected under copyright by the Publisher (other than as may be noted herein).

Notices
Knowledge and best practice in this field are constantly changing. As new research and experience broaden our understanding, changes in research methods, professional practices, or medical treatment may become necessary.

Practitioners and researchers must always rely on their own experience and knowledge in evaluating and using any information, methods, compounds, or experiments described herein. In using such information or methods they should be mindful of their own safety and the safety of others, including parties for whom they have a professional responsibility.

To the fullest extent of the law, neither the Publisher nor the authors, contributors, or editors, assume any liability for any injury and/or damage to persons or property as a matter of products liability, negligence or otherwise, or from any use or operation of any methods, products, instructions, or ideas contained in the material herein.

ISBN: 978-0-12-820709-3
ISSN: 0065-2164

For information on all Academic Press publications
visit our website at https://www.elsevier.com/books-and-journals

Publisher: Zoe Kruze
Acquisitions Editor: Ashlie M. Jackman
Editorial Project Manager: Hilal Johnson
Production Project Manager: James Selvam
Cover Designer: Victoria Pearson

Typeset by SPi Global, India

Contents

Contributors   *vii*

1. **Gaps in the assortment of rapid assays for microorganisms of interest to the dairy industry**   1
   John O'Grady, Ultan Cronin, Joseph Tierney, Anna V. Piterina, Elaine O'Meara, and Martin G. Wilkinson

   1. Dairy industry need for rapid methods, organisms of interest and current rapid methods   2
   2. Organisms for whom no assay or very few rapid assays exist   32
   3. Alternative microbial assay validation   38
   4. Effects of dairy matrices on assay performance   40
   5. Conclusions and recommendations   49
   References   50

2. **The microbiology of red brines**   57
   Aharon Oren

   1. Introduction   58
   2. The organisms   60
   3. Metabolic interactions between the biota   72
   4. The pigments of the red brines   75
   5. Natural hypersaline environments: Case studies   80
   6. Commercial salt production facilities   85
   7. The role of pigmented microorganisms in salt production   96
   8. Concluding remarks   97
   Acknowledgments   99
   References   99

3. ***Clostridium thermocellum*: A microbial platform for high-value chemical production from lignocellulose**   111
   R. Mazzoli and D.G. Olson

   1. Introduction   113
   2. Overview of *C. thermocellum* central metabolism   116
   3. Improving fermentation of hemicellulose-derived sugars by *C. thermocellum*   120
   4. Improving production of high-value chemicals in *C. thermocellum* by metabolic engineering   122

5. Conclusion	147
Acknowledgments	150
Conflict of interest	150
References	150

4. **Predetermined clockwork microbial worlds: Current understanding of aquatic microbial diel response from model systems to complex environments**   **163**
Daichi Morimoto, Sigitas Šulčius, Kento Tominaga, and Takashi Yoshida

1. Introduction	164
2. Culture-based understanding of microbial diel patterns at each trophic level	167
3. Community diel cycles generate from complex biological interactions	173
4. Conclusions and perspectives	180
Acknowledgments	182
References	182

Contributors

Ultan Cronin
Department of Biological Sciences, University of Limerick, Limerick, Ireland

R. Mazzoli
Structural and Functional Biochemistry, Laboratory of Proteomics and Metabolic Engineering of Prokaryotes, Department of Life Sciences and Systems Biology, University of Torino, Torino, Italy

Daichi Morimoto
Graduate School of Agriculture, Kyoto University, Kyoto, Japan

Elaine O'Meara
Department of Biological Sciences, University of Limerick, Limerick, Ireland

John O'Grady
Dairy Processing Technology Centre, University of Limerick, Limerick, Ireland

D.G. Olson
Thayer School of Engineering, Dartmouth College, Hanover, NH; Center for BioEnergy Innovation, Oak Ridge National Laboratory, Oak Ridge, TN, United States

Aharon Oren
The Institute of Life Sciences, The Hebrew University of Jerusalem, Jerusalem, Israel

Anna V. Piterina
Dairy Processing Technology Centre, University of Limerick, Limerick, Ireland

Sigitas Šulčius
Laboratory of Algology and Microbial Ecology, Nature Research Centre, Vilnius, Lithuania

Joseph Tierney
Glanbia Ingredients Ireland, Ballyragget, Co. Kilkenny, Ireland

Kento Tominaga
Graduate School of Agriculture, Kyoto University, Kyoto, Japan

Martin G. Wilkinson
Department of Biological Sciences, University of Limerick, Limerick, Ireland

Takashi Yoshida
Graduate School of Agriculture, Kyoto University, Kyoto, Japan

CHAPTER ONE

Gaps in the assortment of rapid assays for microorganisms of interest to the dairy industry

John O'Grady[a], Ultan Cronin[b,*], Joseph Tierney[c], Anna V. Piterina[a], Elaine O'Meara[b], and Martin G. Wilkinson[b]

[a]Dairy Processing Technology Centre, University of Limerick, Limerick, Ireland
[b]Department of Biological Sciences, University of Limerick, Limerick, Ireland
[c]Glanbia Ingredients Ireland, Ballyragget, Co. Kilkenny, Ireland
*Corresponding author: e-mail address: ultan.cronin@ul.ie

Contents

1. Dairy industry need for rapid methods, organisms of interest and current rapid methods — 2
   1.1 Dairy industry needs and drivers for the application of rapid methods — 2
   1.2 Current detection methods — 11
   1.3 Organisms and groups of organisms of interest to the Irish dairy industry — 15
2. Organisms for whom no assay or very few rapid assays exist — 32
   2.1 Organisms of interest to the dairy industry for which gaps exist in the assortment of rapid assays — 32
   2.2 Organisms/groups of organisms with low coverage of rapid assays — 34
   2.3 Organisms/groups of organisms with no coverage of rapid assays — 35
3. Alternative microbial assay validation — 38
   3.1 Regulations surrounding alternative microbial test methods — 38
   3.2 Paths for adoption of rapid test methods — 38
4. Effects of dairy matrices on assay performance — 40
   4.1 Dairy products for microbiological testing — 40
   4.2 The effect of dairy product varieties on alternative test system performance — 42
   4.3 Strategies to reduce dairy matrix interference — 45
5. Conclusions and recommendations — 49
References — 50

Abstract

This review presents the results of a study into the offering of rapid microbial detection assays to the Irish dairy industry. At the outset, a consultation process was undertaken whereby key stakeholders were asked to compile a list of the key microorganisms of interest to the sector. The resultant list comprises 19 organisms/groups of organisms divided into five categories: single pathogenic species (*Cronobacter sakazakii*, *Escherichia coli* and *Listeria monocytogenes*); genera containing pathogenic species

(*Bacillus, Clostridium, Listeria, Salmonella; Staphylococcus*); broad taxonomic groupings (Coliforms, Enterobacteriaceae, fecal Streptococci, sulfite reducing bacteria/sulfite reducing Clostridia [SRBs/SRCs], yeasts and molds); organisms displaying certain growth preferences or resistance as regards temperature (endospores, psychrotrophs, thermodurics, thermophiles); indicators of quality (total plate count, *Pseudomonas* spp.). A survey of the rapid assays commercially available for the 19 organisms/groups of organisms was conducted. A wide disparity between the number of rapid tests available was found. Four categories were used to summarize the availability of rapid assays per organism/group of organisms: high coverage (>15 assays available); medium coverage (5–15 assays available); low coverage (<5 assays available); no coverage (0 assays available). Generally, species or genera containing pathogens, whose presence is regulated-for, tend to have a good selection of commercially available rapid assays for their detection, whereas groups composed of heterogenous or even undefined genera of mainly spoilage organisms tend to be "low coverage" or "no coverage." Organisms/groups of organisms with "low coverage" by rapid assays include: *Clostridium* spp.; fecal Streptococci; and *Pseudomonas* spp. Those with "no coverage" by rapid assays include: endospores; psychrotrophs; SRB/SRCs; thermodurics; and thermophiles. An important question is: why have manufacturers of rapid microbiological assays failed to respond to the necessity for rapid methods for these organisms/groups of organisms? The review offers explanations, ranging from the technical difficulty involved in detecting as broad a group as the thermodurics, which covers the spores of multiple sporeforming genera as well at least six genera of mesophilic non-sporeformers, to the taxonomically controversial issue as to what constitutes a fecal *Streptococcus* or SRBs/SRCs. We review two problematic areas for assay developers: validation/certification and the nature of dairy food matrices. Development and implementation of rapid alternative test methods for the dairy industry is influenced by regulations relating to both the microbiological quality standards and the criteria alternative methods must meet to qualify as acceptable test methods. However, the gap between the certification of developer's test systems as valid alternative methods in only a handful of representative matrices, and the requirement of dairy industries to verify the performance of alternative test systems in an extensive and diverse range of dairy matrices needs to be bridged before alternative methods can be widely accepted and adopted in the dairy industry. This study concludes that many important dairy matrices have effectively been ignored by assay developers.

1. Dairy industry need for rapid methods, organisms of interest and current rapid methods

1.1 Dairy industry needs and drivers for the application of rapid methods

The key driver for this research, carried out on behalf of the Dairy Processing Technology Centre (DPTC) in conjunction with its eight industrial partners from the Irish dairy industry (Arrabawn, Aurivo, Carbery, Dairygold,

Glanbia, Kerry, Lakeland Dairies and Tipperary Co-op), was to examine currently available commercial tests capable of reducing the time to result (TTR) of microbial analyses in dairy production. TTR is critical for microbiological analyses carried out in a dairy setting: for many products if testing time can be significantly reduced this results in reduced hold/warehousing time and associated costs, as well as quicker time to market for final products. Standard ISO-based plating methods, which have been employed in the industry for over 100 years to evaluate microbial risk via analysis of in-process and finished product test streams, are time consuming, requiring two or more days of incubation prior to reading, with positive results requiring further plating and incubation to confirm putative identifications (Boor, Wiedmann, Murphy, & Alcaine, 2017; Wilkinson, 2018). Further disadvantages are that they are typically labor intensive, incorporate long lead times and produce a large amount of consumable waste.

In the future post-COVID-19 economy, it is key that rapid microbiological testing is an enabler of prompt, informed decision-making in-process and for finished product release. Rapid microbiological methods are key to ensure that process efficiencies are realized in the supply chain. In particular, when producing perishable and/or high-value dairy powders it is key that the food processor achieves the correct first-time metrics related to product release to avoid unnecessary rework and/or product downgrade or disposal. Time to release of product is paramount to ensure that the customer gets the freshest product possible and the Food Business Operator (FBO) holds minimum stock such that warehousing and transport costs can be controlled.

Rapid microbial methods deliver better process control by allowing for prompt intervention when an increase in microbial risk arises within the process chain. Detection of a microbial issue in, for example, 4h allows the operator to intervene and divert the dairy stream through a different processing route or alter processing parameters, rather than continuing to produce a final product that is inevitably out of specification. Timely determination of whether a process is in-control or out-of-control, capable or not capable, based on defined process conditions and run time, is very useful for the process operator. For example, if a customer specification for spores in whey powder is 500 CFU/g and the in-process whey stream contains 800 CFU/g then it is highly likely that the finished powder will be out of specification and, therefore, rapid quality data allows an intervention such as further heat treatment and/or filtration of the liquid stream to achieve correct product specification. Rapid microbiological test methods are essential to improving process efficiencies as well as improving food safety/food spoilage screening and control.

1.1.1 Areas of consideration needed to satisfy ISO 16140 and FBO/customer core criteria

There is a triangular dynamic at work which dictates the nature of microbial testing in the dairy industry, be that conventional or rapid. At the apices of the triangle are the FBO, regulatory authorities (embodied in ISO 16140) and the customer. Each of these parties has their own version of the ideal microbial assay. Regulatory authorities are typically interested in the "business end" of the assay: accuracy; precision; repeatability; reproducibility; specificity; sensitivity; the limit of quantification (LOQ); and the limit of detection (LOD). The cost or administrative burden testing imposes on the FBO are of little interest to regulatory authorities. Customers are similarly interested in the assay's performance and reliability, as this has a direct impact on product quality and shelf-life. However, expensive testing is ultimately included in the price of a product, and so it is in their interest that the cost of testing is as low as possible without compromising on quality or safety. Rapid assays streamline production and logistics for both producer and customer. Below, the characteristics of rapid assays are discussed from the point of view of what is required to meet the needs of the Irish dairy industry. Insights into the ideal assay/instrument/technology were gained through the same process of consultation as described above.

1.1.1.1 Physical characteristics

1.1.1.1.1 Installation difficulty Ideally, any new assay should be implementable on instruments already installed in the plant. For example, a new enzyme-linked immunosorbent assay (ELISA) for an organism that was previously detected using traditional plating, must be readable on the company's existing plate reader. However, if a new assay does require the installation of a new instrument or platform this should be as inexpensive and trouble-free as possible. The ideal platform is plug and play. Platforms which require plumbing, air conditioning, special power supply, or construction work will always be less attractive to laboratory managers.

1.1.1.1.2 Footprint Culture-based techniques require large amounts of space for serial dilutions, inoculation and spreading, as well as for incubation and storage of agar plates and enrichment media. Microbiological testing technologies with a small footprint that reduce the amount of space required to carry out testing procedures allow higher throughput of test samples and redistribution of laboratory space for other purposes. The smaller an instrument or platform's footprint and the smaller the format of testing unit (compare the space occupied by a multiwell plate with 96 Petri dishes) the more attractive it is to a laboratory manager.

1.1.1.2 Operation and use

1.1.1.2.1 Robustness of kit components and instrument/platform End users of assays and instruments are happiest with those that survive the rough-and-tumble of daily laboratory use. Instruments which require regular elaborate calibration or suffer constant breakdowns will not survive on the market. Instruments which block or require long cleans between samples or sample runs will similarly acquire poor reputations.

1.1.1.2.2 Required expertise The more expertise required to execute a method, the higher the pay grade of laboratory personnel required. This contributes to the end cost of testing. Ideally, an assay should be so simple that it can be performed by plant general operatives.

1.1.1.2.3 System suitability A rapid assay, especially those which involve the purchase of an instrument or platform, should be suited to the context into which they are introduced. It should not be too complicated or elaborate for the level of training or expertise of the laboratory personnel. It should not demand more (or less) samples than the laboratory's throughput allows. It should not force the workflow of the laboratory or quality system's testing regime to alter to its preferred workflow, but to fit into existing workflows with minimal fuss.

1.1.1.2.4 Shelf-life Reagents with short shelf lives do not render an assay, system or platform attractive. This complicates ordering, storage and leads to unnecessary waste.

1.1.1.2.5 Refrigeration Reagents which require refrigeration or which need to be stored under special conditions are unwelcome in a laboratory setting, where fridge space is at a premium.

1.1.1.2.6 Hazardous reagents or by-products Assays involving hazardous reagents or by-products introduce complications to do with their storage, handling and disposal, as well as with the training of personnel.

1.1.1.3 Sustainability
The amount of disposable consumables associated with an assay, instrument or platform compared to traditional culture-based detection methods is increasingly an important factor in the decision to implement any new assay.

Culture-based techniques utilize large quantities of plasticware and large volumes of microbiological media that subsequently require autoclaving prior to disposal. Testing methodologies which reduce the quantity of consumable solids and liquids could benefit the companies in terms of reduced costs of and workloads associated with waste disposal, while also allowing companies to reach environmental sustainability targets.

1.1.1.4 Integration with existing systems

1.1.1.4.1 Is multiplex detection possible? Microbiological testing in the dairy environment frequently involves testing of the same product for multiple microorganisms. Using traditional methods this involves separate tests (essentially different media incubated under different conditions) for each microorganism of interest. Novel technologies which allow multiplexing (i.e. detecting multiple organisms of interest in the same sample) would reduce workload, introduce time savings and reduce sample volume, reagents and plasticware. Furthermore, single platforms that allow performance of multiple tests hold the possibility of replacing a number of stand-alone instruments which only carry out testing of a single organism, saving space, reducing workflow complexity and streamlining the testing process.

1.1.1.4.2 Integration with existing workflow A new assay, instrument or platform should not disrupt the existing testing workflow. Ideally, a new assay (and certainly a new platform) should offer the possibility of streamlining the lab's workflow. New assays should not require the purchase of new equipment but should piggy-back on existing equipment.

1.1.1.4.3 Suitable as in-house method In many instances, companies deploy a non-standard assay (perhaps neither ISO/AFNOR-validated nor required by regulatory authorities) for in-house process monitoring or quality assurance. If this assay is based on traditional methodology, the company may be in the market for a rapid-format assay to replace it.

1.1.1.4.4 Compatibility with laboratory information management system (LIMS) Industrial testing procedures from sources such as raw materials, intermediates, end products and production environment are integrated with processing and plant management systems via computer databases. Considerations of novel testing platforms include their compatibility with existing data management and integration systems, as well as existing workflows, such as sampling workflows.

1.1.1.5 Cost

1.1.1.5.1 Cost of capital equipment Instrumentation associated with assays can be expensive, especially if an entire platform requires purchasing. Companies would obviously like to pay as little as possible to introduce a new rapid assay into their laboratory. Capital costs, running costs, staff costs, etc. should be considered together to give an overall cost per sample. On some occasions a platform with high capital costs may be associated with low running costs, and so, over time lead to savings in testing. A detailed cost-benefit analysis may indicate that investment in a new platform will eventually lead to savings in e.g. dropped batches, warehousing, recalls.

1.1.1.5.2 Running cost While dairy companies would like to reduce the cost of testing per sample with new technologies, or at least to maintain the cost of testing at the same level as current standard methods, the effect that the new technologies have in terms of TTR, throughput, reduction in footprint and labor h, for example, could allow more expensive rapid technologies to be considered. The total cost of assay platforms is based on the individual cost of the capital equipment and running costs, including consumables and equipment maintenance. Volume of reagent use, maintenance contracts with manufacturers, and the cost of spare parts should all be taken into account when calculating running costs.

1.1.1.6 Validation and certification

Whether a method is validated for the dairy matrix of interest to the FBO and certified by one of the bodies described in Section 3.2 is an important factor in a company's adapting a new assay. An assay vendor's claims for a kit may fail to impress if not backed up by the guarantee of its validation for the task for which it is intended. FBOs very much see validation as the work of assay/platform/instrument manufacturers: even the largest FBO may not have the personnel or expertise to validate an assay. As an extra seal of quality, if an assay has been certified it is a demonstration to a potential FBO customer that they are investing in a quality product which has been independently demonstrated to work.

1.1.1.7 Sample/matrix

The constituents of food matrices can affect the performance of rapid microbiological technologies, and some matrices may not be compatible with certain technologies, or only after sample pre-treatment to reduce or remove the interfering matrix components is performed (see Section 4).

Furthermore, the range of dairy products and ingredients being produced, and the inclusion of intermediates as well as raw and final products for testing, creates an extensive portfolio of matrices of varying fat, whey, casein and lactose compositions. Therefore, prior demonstration of the compatibility of the rapid technology with all matrix compositions of interest to produce accurate results is of vital importance to food industries before a rapid technology is adopted.

1.1.1.8 Enrichment time

The ideal assay would enable immediate and direct detection of the microorganism of interest. However, the microorganism of interest is often present in numbers below the limit of detection of the assay and so the sample requires enrichment in order to allow the organisms to grow to numbers capable of being detected by the method. It goes without saying that extended enrichment times set alarm bells ringing for any laboratory manager looking to source a commercial rapid assay.

1.1.1.9 Qualitative or quantitative

Depending on the organism or group of organisms to be detected, an FBO may be interested either in its qualitative or quantitative detection. For some organisms, the company's hands are directed by regulatory demands, but where the freedom to choose exists whether an assay is qualitative or quantitative, the decision very much depends on what use the data will serve. Initial screenings may often be "presence/absence." Should the organism of concern be present then a subsequent quantitative assay will be deployed.

1.1.1.10 Technical performance

1.1.1.10.1 Accuracy Along with rapidity, a high degree of accuracy is the most desirable technical attribute of an assay (Hameed, Xiea, & Ying, 2018). Accuracy is the correspondence between the result generated by the assay and the biological reality—how many microbes did the assay detect versus how many bacteria were actually present. This figure is determined statistically and holds only within a range of target bacteria and for a given matrix.

1.1.1.10.2 Precision Sound methods are associated with high precision, i.e. agreement between repeated measurements. Statistical data on precision should be provided by assay vendors, especially in the case of novel technologies, as these are often associated with low precision (Wei, Wang, Sun, & Pu, 2019).

1.1.1.10.3 Repeatability Repeatability is of interest to FBOs to enable comparison of data generated by an assay over time (Anon, 2019). Methods (or instruments or platforms) associated with poor repeatability, where results jump wildly from day to day, or show drift, are unacceptable, as trends such as increasing contamination of raw milk during droughts may not be detected.

1.1.1.10.4 Reproducibility This value expresses the precision between laboratories (Anon, 2019) and is of interest to FBOs which conduct testing over multiple sites. Poor reproducibility could be a result of different personnel, equipment, laboratory practices or different assay platform instruments being used in different plants.

1.1.1.10.5 Specificity Methods with low specificity have high false positive rates (Anon, 2019; Hameed et al., 2018). False positives are costly for the dairy industry. At the very least they lead to further testing, while they could wrongly lead to delayed release of batches, the downgrading of a powder's quality, or an extra processing step to deal with the "out of spec" reading.

1.1.1.10.6 Sensitivity One of the most desirable technical attributes of a microbial detection assay, sensitivity, in, layman's terms is the lowest level of a microorganism that can accurately be measured by a method. It reflects the ability of a method to "find" all of the target organisms in a sample, i.e. detect the true positives (Hervert, Alles, Martin, Boor, & Wiedmann, 2016). Methods which lack sensitivity will not find all of the target organisms present in a sample, leading to false negative results. These are highly unwelcome in an industry where recalls can cause severe economic and reputational damage.

1.1.1.10.7 LOQ This is the lowest number of microbes that can be quantitatively determined with an acceptable level of uncertainty (Anon, 2019). At the very least, the LOQ of an assay must be capable of quantifying CFU/mL within the range specified by legislation. FBOs would desire that the LOQ be an order of magnitude below this, if possible, in order to monitor raw materials and products for quality and take preventative action before compliance is breached.

1.1.1.10.8 LOD The lowest amount of analyte in a sample which can be detected but, not necessarily quantified, as an exact value (Anon, 2019). If an assay's LOD is significantly below the LOQ an assay could be used as outlined above—with qualitative data indicating that a product or process is approaching being out of tolerance.

1.1.1.10.9 Throughput There is a requirement that instruments and platforms for microbial detection assays in the dairy industry be high throughput, with figures of "100s" of samples/h commonly mentioned. High throughput implies limited manual processing of samples and high degrees of automation. Such systems, however, tend to be associated with higher capital costs.

1.1.1.10.10 Linearity and range An assay which is capable of delivering accurate results over the range required by dairy FBOs meets their demands perfectly. In the case of assays for certain microbes, a short range may be adequate to meet the needs of the client, while for other microbes, producers may be interested in enumerating over a number of log decades.

1.1.1.11 Manufacturer support
1.1.1.11.1 Training/set-up One of the key factors in the decision to go with an assay, instrument or platform is the extent and quality of the training provided by the manufacturer to the FBO's key personnel. A common complaint during consultations with the DPTC's industrial partners for the preparation of this review was that reagent and instrument manufacturers were very willing to sell an assay, but very reluctant (probably because of the cost involved) to provide training and assistance for the setting up of the assay in the FBO's own labs. If vendors invest time and effort in setting up the assay in the company's facility so that the integration of the assay into the FBO is seamless, this is a considerable selling point for that assay. Often the expertise does not exist in an FBO for the setting up of a new assay involving a new technology. For example, a laboratory which has always relied upon traditional culture techniques (plating and most probable numbers [MPN]) will not have the necessary experience to implement a PCR-based assay without outside assistance.

1.1.1.11.2 Service offering Another common complaint of FBOs is the poverty of service offered by reagent and instrument vendors post sale. Laboratory managers want the surety of knowing that if a complex piece of technology is giving problems or performing sub-optimally that the vendor will solve this matter.

1.1.1.11.3 Manufacturer response time In order to reduce down-time of the testing procedure, a rapid response by manufacturers to technical issues that arise with instruments is a must. FBO's ideally wish for same-day responses to issues with instruments, and demand such services as 24 h hotlines for technical support and troubleshooting.

1.2 Current detection methods

The information on the currently available assays/instruments/platforms was gathered as described in Section 1.1 (through contact with the dairy industry partners), as well as from literature searches and direct contact with assay, instrument and platform sales and technical personnel. All assays, instruments and platforms found at the time of writing (May 2020) were commercially available. Where possible, all claims of validation were verified through organizations such as AFNOR, AOAC, NordVal, etc. In some cases, verification of validation could not be confirmed, as the information was not available.

Forty-one commercial test systems (stand-alone units capable of performing a specified range of microbiological assays) were found that could be applied to the organisms of interest to the Irish dairy industry and which were claimed to have been tested on these organisms in dairy matrices by their manufacturers. A wide variety of analytical techniques or platforms were represented, including flow cytometry (FCM), quantitative polymerase chain reaction (PCR), ELISA, enzyme-linked fluorescence assay (ELFA), matrix-assisted laser desorption/ionization-time of flight (MALDI-TOF), MPN, spectroscopy, lateral flow, impedance, bioluminescence and chromogenic media. Below is a short description of the principal technologies underlying the rapid assays.

1.2.1 FCM

FCM is a method based on the flow of single cells past one or more lasers (Kennedy & Wilkinson, 2017). Scatter and fluorescence data are recorded from thousands of individual cells per second. The method has been used for almost 20 years to enumerate bacterial cells in dairy samples (Gunasekera, Attfield, & Veal, 2000). It is fast, accurate and widely used in the dairy industry to generate counts analogous to total plate counts. Four platforms use FCM to analyze samples: bioMérieux's D count (bioMérieux, Marcy-l'Étoile, France), FOSS's BactoScan™ and BacSomatic™ (Foss, Hilleroed, Denmark), Bently Instruments' BactoCount (Bently Instruments, Chaska, MN, USA) and Sigrist's BactoSense (Sigrist, Ennetbürgen, Switzerland). Sigrist's system, while to date is not being used to enumerate bacteria in dairy products, is used to determine on-line continuous cell counts in water samples, and as such can be used by the dairy industry for water and CIP rinsate analysis. FCM has the potential to do much more in a dairy setting than perform total counts. Combined with antibody tagging and physiological staining the technology has the potential to become the basis of a multiplexed detection system (up to a dozen antibodies and stains can be simultaneously applied) which will detect live, dead or vital microbes down to strain level (Kennedy & Wilkinson, 2017).

1.2.2 Quantitative PCR

Quantitative PCR is a method based on amplifying specific DNA sequences belonging to a target species, genus or broader taxonomical group, and, through measuring the increase in concentration of the target sequence over time and comparing this with a standard curve, quantifying the initial number of genomes present in the sample (Hameed et al., 2018). The technique is rapid and among the most sensitive detection methods. However dairy industry users hesitate to use it because of its technical difficulty, problems with repeatability and expense. There are also complications in relating detected genome copy numbers to the numbers of viable microorganisms (McHugh, Feehily, Hill, & Cotter, 2017), as well as food matrix interference (see Section 4). Ten platforms use PCR to analyze samples; ThermoFisher SureTect™ (ThermoFisher Scientific, Waltham, MA, USA), ThermoFisher MicroSEQ, bioMérieux GENE-UP®, bioMérieux Invisible Sentinel, bio-rad iQ Check (Bio-rad, Hercules, CA, USA), Hygiena BAX Q7 (Hygiena, Wilmington, USA), Pall GeneDisc (Pall, Port Washington, NY, USA), BIOTECON foodproof® (BIOTECON, Potsdam, Germany) and Merck Assurance GDS (Merck, Kenilworth, NJ, USA).

1.2.3 ELISA

This is the most commonly used immunoassay for the detection of bacteria, and often regarded as the most convenient (Ziyaina, Rasco, & Sablani, 2020). This is generally employed in a multiwell plate format and is amenable to automation and high throughput. ELISAs can be quantitative or qualitative, in the latter case with a simple color change indicating the presence or absence of the microbe being assayed for. As is the case with other immunological methods, ELISAs work on the basis of labeled antibodies that bind to antigens on the cell surface of specific microorganisms or groups of organisms (Poghossian, Geissler, & Schöning, 2019). The development of further ELISAs for the dairy industry depends on the availability of highly specific antibodies against the microbes of interest. This is also the case with other immunological methods—FCM and ELFA. One platform uses ELISA as an analytical method: Solus Scientific's Solus One (Solus Scientific, Mansfield, UK).

1.2.4 ELFA
This technique is related to the ELISA, with the difference being that for ELFA the antigen is tagged with a bioluminescence-generating molecule such as luciferin, which emits light when oxidized by the enzyme luciferase, rendering the method more sensitive than the ELISA (Hameed et al., 2018). One commercial platform uses ELFA: bioMérieux's VIDAS®.

1.2.5 MALDI-TOF
MALDI-TOF is a mass spectroscopy-based technology which allows analytical determination of biomolecules by creating and detecting ions. The particular patterns created by the ionization of a microbial species' cell can be used to identify the organism in samples (Hameed et al., 2018). While this is a costly platform, it offers the advantage of multiplexing and the addition of further detection assays going forward. One platform uses MALDI-TOF: bioMérieux's VITEK® MS.

1.2.6 MPN
The MPN method has been in existence since the very beginning of analytical microbiology (Boor et al., 2017). In terms of rapid methods, though, what is in question is the miniaturization of the traditional MPN method, coupled with sensitive measurement technology which allows the detection of growth much sooner than in the bulk assay (Sohier, Pavan, Riou, Combrisson, & Postollec, 2014). Being growth based, the results of this type of assay can never be real-time, as other techniques such as FCM have the potential to be. Miniaturized MPN has the potential to be automatable, high throughput and simple to perform and interpret results. Two platforms use MPN methods: bioMérieux's TEMPO® and SY-LAB's AMP 6000® (SY-LAB, Neupurkersdorf, Austria).

1.2.7 Spectroscopy
There are many "flavors" of spectroscopic techniques, which all involve interactions between matter and electromagnetic radiation: hyperspectral imaging, fluorescence spectroscopy, UV–visible, near infrared (NIR), mid infrared (MIR) and far infrared.

(FIR) spectroscopies, and Fourier transform infrared spectroscopy (FTIR; Hameed et al., 2018; Ziyaina et al., 2020). FTIR, Raman spectroscopy, and hyperspectral imaging show the most potential as methods of detecting microbes in foodstuffs. One platform uses spectroscopy: Neogen's Soleris® (Neogen, Lansing, MI, USA).

1.2.8 Lateral flow
The lateral flow assay is a type of immunoassay where the sample travels along a pad via capillary action to the test and control lines (Hameed et al., 2018). If the sample contains the organism of interest both lines darken. Lateral flow assays are simple to use and give clear presence/absence results, which are generally instantaneous. These assays are very convenient for plant hygiene monitoring or field testing and require little laboratory infrastructure. They can be used as pre-screening methods, where positive samples are further investigated using quantitative techniques. Four platforms use lateral flow: Romerlab's RapidChek® (Romerlab, Getzersdorf, Austria), Merck-Milipore's Singlepath®, SY-LAB's RiboFlow® and Neogen's REVEAL® 2.0.

1.2.9 Impedance
This technology is based on measuring changes in a solution's impedance which reflect the growth and metabolism of the organism of interest (Ziyaina et al., 2020). Electrochemical methods are cost-effective but they have low sensitivity and limited selectivity compared to optical methods (Wei et al., 2019). Two approaches can be used in the impedance-based detection of microorganisms: the use of selective medium with impedance measurement; the combination of impedance with biological recognition technology. Since the latter does not rely on growth, it has the potential to be a near real-time method. Two platforms use impedance as a method of microbial detection: SY-LAB BacTrac 4300 & SY-LAB BioTrac 4250.

1.2.10 Bioluminescence
In the ATP bioluminescence technique, the enzyme luciferase catalyzes a chemical reaction with ATP and luciferin to generate light (Ziyaina et al., 2020). The amount of light generated can be related to the number of ATP molecules present, or, indeed the number of cells present. Although the technology is best known for monitoring plant hygiene, it can be used for the detection of contaminating microorganisms, with its speed and sensitivity comparable to that of other technologies (Poghossian et al., 2019).

Six platforms use bioluminescence: Merck Milliflex®, Merck MVP Icon, Merck HY-Lite® 2, Hygiena EnSURE™ Touch, 3M™ Clean-Trace™ (3M, St. Paul, MN, USA) and r-biopharm's Lumitester PD-30 (Darmstadt, Germany).

1.2.11 Chromogenic media
The development of chromogenic media, which allow clear identification of microbial species or groups, constituted a significant breakthrough in the detection of microorganisms of interest to the dairy industry (Boor et al., 2017). Even though chromogenic media require growth to colony or micro-colony level, they do constitute a rapid technology as steps such as enrichment or further testing of putative positive colonies are eliminated (Sohier et al., 2014). Five chromogenic media platforms for dairy monitoring were identified: Liofilchem Contam Swab (Liofilchem, Waltham, MA, USA), bio-rad Rapid Medium, CHROMagar (Chromagar, Springfield, NJ, USA), ThermoFisher Brilliance and Biokar COMPASS (Biokar Diagnostics, Paris, France).

1.3 Organisms and groups of organisms of interest to the Irish dairy industry
1.3.1 Consultation with industry
During the latter half of 2019, an initiative was undertaken whereby the DPTC's eight industry partners were engaged in a process of consultation with the authors, the purpose of which was to identify gaps in the offering of rapid assays by commercial enterprises for organisms or groups of organisms of interest to the Irish dairy sector. Over the course of a number of meetings, a list of relevant organisms was compiled. This list is complex, with 19 organisms or groups thereof, and may contain some entries (and indeed some omissions) which may come as a surprise to academic researchers in the field of dairy microbiology, but not to those "on the ground" in the dairy industry (Table 1). The absence of *Campylobacter jejuni*, which in 2017 featured as number seven in the top 10 pathogen/food vehicle pair causing the highest number food-borne outbreaks in the EU (EFSA & ECDC, 2018), may constitute one such omission. However, it must be understood that organisms such as *C. jejuni*, which make up the natural microflora of cow's milk but which are eliminated by pasteurization and do not reappear to contaminate products during downstream processing are not of interest to the dairy industry (Artursson, Schelin, Lambertz, Hansson, & Engvall, 2018). On the other hand, the Irish dairy industry's

Table 1 The microorganisms of interest to the Irish dairy industry and the reasons for testing for them.

Organism/group	Regulations	Raw material quality	Hygiene	Biofilms	Powder	Shelf life	Post-pasteurization contamination	Infant formula
Bacillus spp.	Y	Y			Y			Y
Clostridium spp.	Y				Y			Y
Coliforms		Y	Y				Y	
Cronobacter sakazakii	Y				Y			Y
Endospores		Y		Y	Y			Y
Enterobacteriaceae	Y	Y	Y				Y	
Escherichia coli	Y							
Fecal streptococci			Y	Y				
Listeria spp.	Y			Y				
L. monocytogenes	Y			Y				
Pseudomonas spp.			Y	Y		Y	Y	
Psychrotrophs		Y		Y		Y		
Salmonella spp.	Y							
SRBs/SRCs			Y		Y			Y
Staphylococcus	Y							
Thermodurics			Y	Y	Y			Y
Thermophiles				Y	Y	Y		Y
Total plate count	Y	Y	Y			Y		
Yeasts and molds			Y	Y		Y		

strong interest in the detection of fecal Streptococci may not be reflected in other jurisdictions. It is chiefly the *Enterococcus* spp. component of the fecal Streptococci, which are known to cause the spoilage of cheese, which is the Irish dairy industry's concern.

Two concepts govern the appearance of an organism/group of organisms on the list: compliance and process control. If an organism's detection is necessary for regulatory compliance it is on the list. Additionally, if an organism's (or usually group of organism's) detection provides a useful insight into the dairy manufacturer's process then it is also on the list. The list may be divided into five categories: single species which are pathogens; genera containing pathogenic species; broad taxonomic groupings; organisms displaying certain growth preferences or resistance as regards temperature; indicators of quality. From a food safety standpoint, the importance of many of the organisms included on the list cannot be disputed (7.7% of foodborne outbreaks in the EU in 2017 involved dairy produce; EFSA & ECDC, 2018): this would include all the species or genera to which pathogens belong and all of which fall under stern regulatory criteria regarding their levels in finished product. Beyond the detection of pathogens necessary for the hygienic control of dairy products and their regulatory compliance, dairy producers have an interest in the detection of microbes or groups thereof which provide them with data on raw material quality, CIP and the operation of equipment such as pasteurizers, spray driers and ultrafiltration plants: many of the broad taxonomic groupings, organisms displaying certain growth preferences or resistance as regards temperature and indicators of quality are of interest to dairy processors precisely for this reason. Knowledge of their numbers at a given point in time in a process, as well as trends in the evolution of these numbers, provides increased control over and confidence in a particular process or stage in that process.

From the point of view of the microbial taxonomist it may be anathema to express an interest in detecting e.g. sulfite-reducing Clostridia (SRCs) or coliforms, as these groupings are often disputed in the literature and/or contain widely unrelated species. Even more puzzling to the taxonomist may be the desire to detect the best part of an entire kingdom (yeasts and molds) or a tranche of unrelated microbes which happen to grow or survive at a certain temperature (e.g. psychrotrophs). Regardless of these groupings' spurious systemic basis, however, they are of proven practical value in that they allow personnel in decision making and planning roles in dairy plants to manage their processes in accordance with the principals of quality assurance and GMP. Companies to which dairy plants supply materials for input into their own product lines (e.g. infant formula manufacturers) often use criteria such as endospore or SRC count as a basis for lot acceptance. Countries outside of the EU regulatory framework may also have stipulations for minimum e.g. endospore counts.

1.3.2 Single species which are pathogens
1.3.2.1 *Cronobacter sakazakii*
Cronobacter sakazakii, along with *Listeria monocytogenes* and *Salmonella* spp., is one of the "big three" pathogens of current concern for postprocessing contamination of dairy products (Boor et al., 2017). Because of the danger it poses to neonates, *C. sakazakii* is seen as a particular risk in infant formula, with Commission Regulation (EC) No. 2073/2005 stipulating its absence in 10 g of dried infant formula and dried dietary foods for special medical purposes intended for infants below 6 months of age (European Commission, 2005). This puts considerable pressure on milk powder manufacturers in terms of the ability to supply a consistently clean product as well as deploying within the boundaries of the Regulation a highly sensitive assay for detection of the organism. *C. sakazakii* finds its way into product through its persistence in the processing environment, where moisture and organic matter favor its survival (Flint et al., 2020). Because of this, along with detection in intermediates and finished product, manufacturers are interested in controlling for *C. sakazakii*'s presence in the plant, meaning the requirement for a convenient and rapid assay (though not necessarily as sensitive as that required for product testing) is urgent. An EU-wide survey found that at retail level, out 1014 samples tested one was reported positive, while at processing plant level, out of 387 samples tested 16 were positive (EFSA & ECDC, 2018).

1.3.2.2 *Escherichia coli*
The testing for *E. coli* in dairy plants, with the exception of those that sell raw milk or produce cheese made from unpasteurized or low-temperature-treated milk is primarily for the purposes of hygiene monitoring (Artursson et al., 2018). Following the decision of the EU's Scientific Committee on Veterinary Measures relating to Public Health that verotoxigenic *E. coli* (VTEC) represented a hazard to public health in raw milk and raw milk products (European Commission, 2005), it is stipulated that these strains be absent in 25 mL of raw milk to be sold as such (Commission Regulation [EC] No 1441/2007; European Commission, 2007). The requirement for cheeses made from milk or whey that has undergone heat treatment to contain no more than 100–1000 CFU/g and butter and cream made from raw milk or

milk that has undergone a lower heat treatment than pasteurization to contain no more than 10–100 CFU/g is described in a footnote of the legislation as an "indicator for the level of hygiene". Notwithstanding concern regarding shiga toxin–producing *E. coli* (STEC), detection of *E. coli* in the dairy industry is primarily with the aim in mind of identifying sources of fecal contamination at points along the production process (Boor et al., 2017; EFSA & ECDC, 2018). As a member of the Enterobacteriaceae, which cycle between mammalian guts and the soil, *E. coli* has long been used as an indicator of fecal contamination. It is especially useful to monitor for *E. coli* and other Enterobacteriaceae in post-pasteurization contexts, given that these do not survive the process and so their presence must be down to fecal contamination through one or more routes (Flint et al., 2020). As an aside, the use of PCR to detect STEC has brought dramatic improvements, with dramatic increases in sensitivity and specificity over the plate-based method was one of the first "killer apps" of PCR in dairy microbiology (Willis et al., 2018).

1.3.2.3 L. monocytogenes

L. monocytogenes is a problematic microorganism for the dairy industry, arguably the most important pathogen associated with this sector (Boor et al., 2017). With the ability to cause severe illness in the elderly, infirm and immunocompromised, as well as abortions, in 2017 there were 2480 cases of invasive listeriosis in the EU, a figure which has seen an increase over the past 5 years of available reports (EFSA & ECDC, 2018). Found in raw milk, and, in spite of being removed by pasteurization, it is capable of finding its way into downstream products through its ability to form biofilms and withstand desiccation (Flint et al., 2020). Strict limits are in place concerning allowable levels in final products. The legal limit is absence in 25 g of product for the majority of dairy products (European Commission, 2005). Raw milk ready to be placed on the market and during its shelf-life has a limit of 100 CFU/mL. It could be stated that *L. monocytogenes* is the scourge of cheeses produced from raw or low-heat-treated milk: a recent extensive study found that up to 3.5% of such semi-soft cheeses and 2.2% of hard cheeses tested positive for the organism (EFSA & ECDC, 2018). Stringent testing of plant for the presence of *L. monocytogenes* is necessary for its control (Boor et al., 2017), something which explains the constant demand among dairy industry quality assurance managers for more sophisticated and rapid detection assays. Indeed, the first PCR-based assay for a dairy microorganism was for *L. monocytogenes*, with a raft of such novel assays appearing in the early 1990s (see Boor et al., 2017). A sophisticated tracing system of all clinical cases of listeriosis has been in place in the US since 1999, with a system called PulseNet International working at a global level. As sequencing becomes more affordable, exquisitely accurate relating of strain to outbreak will become routine.

1.3.3 Genera containing pathogenic species
1.3.3.1 Bacillus
It is not just this genus' foodborne pathogen, *B. cereus*, which is of interest to the dairy industry, but many other problematic spoilage organisms within the genus, such as *B. licheniformis*. Species within the genus, *Bacillus*, being naturally present in the soil occur in high numbers in milk, and, because they are sporeformers, survive pasteurization and other thermal processes, finding their way into finished products such as milk powder or whey protein (Reineke & Mathys, 2020). Endospores of *B. licheniformis*, for example only suffer a 0.01 log reduction in milk following pasteurization (Khanal, Anan, & Muthukumarappan, 2014) and members of the genus can grow as biofilms where both endospores and vegetative cells are present (Park, Yang, Choi, & Kim, 2017). While methods do not have to be as sensitive to detect *B. cereus* and other pathogenic *Bacillus* spp. as they need to be for e.g. *L. monocytogenes* or *Salmonella* in order to comply with regulations ("satisfactory" levels must be below 1.0×10^3 CFU/mL and "unsatisfactory" 1.0×10^5 CFU/mL; FSAI, 2019), customers' specifications for powdered products may be much more stringent (<10 CFU/mL), with the threshold in dried infant formula intended for infants below 6 months of age being 50 CFU/g (Commission Regulation [EC] No. 1771/2007 [European Commission, 2007]). Therefore, detection methods for either *B. cereus* alone or members of the genus in general need to be capable of detecting single endospores per gram of product. Significant testing of raw materials, intermediates and products is carried out in dairy companies to control for members of this genus. Detection methods must ideally be capable of enumerating both endospores and vegetative cells, and, if possible, yield counts for each cell type: such counts would provide valuable data on such things as resistance profiles of a sample and germination potential.

1.3.3.2 Clostridium
As a genus of anaerobic sporeformers, *Clostridium* has not been the subject of as much attention as *Bacillus*, despite the fact that members of the genus occur in raw milk, survive pasteurization and are detected in downstream

products (Reineke & Mathys, 2020). The genus, *Clostridium*, boasts members which are toxigenic, neurotoxigenic or spoilage bacteria (Doyle et al., 2015). The most important species from a dairy microbiology point of view are *C. perfringens* (which causes food poisoning), *C. sporogenes* (a spoiler of cheese) and a subgroup of spoilage bacteria associated with cheese known as the butyric acid bacteria (BAB), which includes the species, *C. butyricum*, *C. tyrobutyricum* and *C. beijerinckii* (Brändle, Domig, & Kneifel, 2016). Additionally, *C. botulinum*, the agent of botulism, is found sporadically in powdered dairy products and is a significant concern for the dairy industry and consumers. There is significant difficulty in detecting members of the genus, as well as designating a specific strain to isolates (Janganan et al., 2016). Uncertainty in the industry regarding which species are problematic, or which species constitute indicators of hygiene lapses is reflected in the desire to detect either (or both) *Clostridium* spp. and SRCs (see below; Doyle et al., 2015). With only the presence of *C. perfringens* ordained to be tested for in the legislation, with levels below 1.0×10^1 CFU/mL deemed "satisfactory" in ready-to-eat foods (FSAI, 2019), it is up to individual companies and their customers which of the other organisms within the genus are to be tested for, or whether it is sufficient to test for the presence of "Clostridia" in general.

1.3.3.3 *Listeria*

Legislation specifies in certain instances the absence of *Listeria* spp. in 25 g of final product (Commission Regulation [EC] No. 2073/2005 [European Commission, 2005]). Additionally, as an indicator of hygiene, legislation states that foods which cannot support the growth of *Listeria* spp. have less than 1.0×10^1 CFU/mL to be considered "satisfactory," while those that can support growth should demonstrate freedom from members of the genus. As indicators of general plant hygiene *Listeria* spp. are an excellent choice, displaying as they do a general hardiness and recalcitrance to CIP only matched by the sporeformers. As well as being able to form biofilms, *Listeria* spp., are psychrotrophs, capable of growth at refrigeration temperatures (Melo, Andrew, & Faleiro, 2015). If a company's hygiene program has succeeded in removing *Listeria* spp. from its plant and processing equipment this is a good indicator that general hygiene is excellent (Boor et al., 2017). If, however, *Listeria* spp. are detected in a certain e.g. ultrafiltration cabinet or storage vessel this is a sign for remedial action to be taken.

1.3.3.4 Salmonella

While there are over 2500 *Salmonella* serovars, the top five most commonly detected serovars in the EU are *S. infantis*, *S. typhimurium*, *S. enteritidis*, monophasic *S. typhimurium* and *S. newport* (EFSA & ECDC, 2018). Even though poultry products and meat are associated in the public mind with the pathogen, dairy products are an important source of *Salmonella* outbreaks: in 2017 1.9% of these were caused by cheese and 1.5% by dairy products other than cheese. Regulations stipulate the organism's absence in 25 g (or mL) of milk and whey powder, cheeses, butter and cream made from raw milk or milk that has undergone a lower heat treatment than pasteurization, and raw milk at point of sale (Commission and Parliament Regulation [EC] No 178/2002, European Parliament and Council, 2002; Commission Regulation [EC] No 1441/2007 [European Commission, 2007]). *Salmonella* is a concern for the post-pasteurization contamination of dairy products (Boor et al., 2017), and, as such, detection and monitoring programs should form a part of a company's quality assurance plan. Given the severity of salmonellosis in the case of some individuals and the damage association with an outbreak can cause to a company, it is important that any rapid method for *Salmonella* detection are sensitive and capable of detection, if possible, all serovars (McClelland & Pinder, 1994).

1.3.3.5 Staphylococcus

When the dairy industry speaks of *Staphylococcus* detection, what they are referring to is a group of the "coagulase-positive Staphylococci" (CPS), the detection of which is stipulated in European legislation. Within the grouping CPS is the prime member, *S. aureus*, as well as *S. intermedius* and *S. hyicus* (Roberson, Fox, Hancock, & Besser, 1992). Along with non-*Staphylococcus* species such as *E. coli*, *Streptococcus agalactiae* and *St. uberis*, the CPS cause bovine mastitis, entering the dairy supply chain through the milk of infected cattle (Vanderhaeghen et al., 2015). Because the disease risk associated with *S. aureus* (and possibly the other CPS) is by virtue of the heat-stable toxins they produce (Islam et al., 2018), regulations can either refer to the detection of toxin or the organisms that produce it. In cheeses, milk powder and whey powder, Commission Regulation (EC) No 1441/2007 stipulate that no enterotoxins should be detected in 25 g of product (European Commission, 2007). Limits for the toxin-producing organisms are less severe: cheeses made from raw milk may contain up to $1.0 \times 10^{4-5}$ CFU/g of CPS; cheeses made from milk that have undergone a lower heat treatment than pasteurization and ripened cheeses made from milk or whey that has undergone pasteurization or a stronger heat treatment may contain up to $1.0 \times 10^{2-3}$ CFU/g of CPS; unripened soft cheeses (fresh cheeses) made from milk or whey that has undergone pasteurization or a stronger heat treatment may contain up to $1.0 \times 10^{2-3}$ CFU/g of CPS; milk powder and whey powder may contain up to $1.0 \times 10^{1-2}$ CFU/g of CPS. Primary legislation states that if greater than 1.0×10^{5} CFU/g of CPS are found in raw milk it has to be tested for staphylococcal enterotoxins (Commission and Parliament Regulation [EC] No 178/2002, European Parliament and Council, 2002). If enterotoxin is detected, the raw milk is considered unsafe. *Staphylococcus* spp. are important components of dairy plant biofilms (Flint et al., 2020) and as such should be a target organism of plant hygiene monitoring programs. The ability of assays to detect all members of the CPS should be verified. A future trend in this area may be the inclusion of coagulase-negative Staphylococci in the species of interest, as these have been implicated in cases of sub-clinical mastitis (Vanderhaeghen et al., 2015).

1.3.4 Broad taxonomic groupings
1.3.4.1 Coliforms
There is no regulatory framework for levels of coliforms in dairy products. Testing for this diverse group (containing 19 genera) of aerobic or facultatively anaerobic, Gram-negative, non-sporeforming rods capable of fermenting lactose to produce gas and acid within 48h at 32–35°C is common practice within the dairy industry as part of its in-house hygiene monitoring (Hervert et al., 2016). The test for coliforms as indicator organisms of fecal contamination of milk was one of the first microbiological assays implemented in the dairy industry almost 100 years ago (Boor et al., 2017). There are many who see the test as having outlived its usefulness, with the only benefit of continued testing being to have an unbroken sequence of data with months and years of previous tests. While initially thought to be a homogenous group of enteric bacteria the presence of which post-pasteurization was evidence of fecal contamination from plant or operator, it has been shown that the majority of coliforms originate the environment and that their presence in milk and dairy products rarely indicates actual fecal contamination (Martin, Trmčić, Hsieh, Boor, & Wiedmann, 2016). It has also been established that coliforms account for only 7.6% to 26.6% of bacteria introduced into fluid milk by post-pasteurization contamination (Martin, Carey, Murphy, Wiedmann, & Boor, 2012). Furthermore, *Pseudomonas* spp., which have been shown to represent the majority of

postprocessing contaminants in fluid milk are not detected by the coliform assay (Sørhaug & Stepaniak, 1997). Many authors point out the superiority of testing for Enterobacteriaceae as markers of post-pasteurization contamination (see below; Hervert et al., 2016).

1.3.4.2 Enterobacteriaceae

In contrast to testing for coliforms, testing for the family, Enterobacteriaceae, as well as providing the dairy manufacturer with an insight into plant and product hygiene, is required for regulatory purposes. Pasteurized milk and other pasteurized liquid dairy products must not contain more than 10 CFU/mL of these organisms (Commission Regulation [EU] No 365/2010, European Commission, 2010), milk and whey powder less than 10 CFU/g, while they must be absent from 10 g of dried infant formulae and dried dietary foods for special medical purposes intended for infants below 6 months of age (Commission Regulation [EC] No 1441/2007, European Commission, 2007). Raw milk must not contain more than 100 CFU/mL of the organisms (Commission and Parliament Regulation [EC] No 178/2002, European Parliament and Council, 2002). The argument that testing for Enterobacteriaceae provides a superior insight into fecal contamination of processed product compared to testing for coliforms is that the former tests for a wider range of organism than the latter, including the important disease-causing genera *Salmonella* and *Yersinia* (Boor et al., 2017; Martin et al., 2016).

1.3.4.3 Fecal streptococci

Many readers of this review may be puzzled by the term "fecal Streptococci" and question why the Irish dairy industry regards detection of this group of organisms as important, when the presence of its members in dairy products are not regulated for, and when the very genera and species comprising the group are still a matter of controversy and confusion. It is worth quoting Franz, Stiles, Schleifer, and Holzapfel (2003) *en bloc* at this point:

> Members of the genus Streptococcus *that were formerly grouped as 'fecal streptococci or Lancefield's group D Streptococci' were subdivided into three separate genera:* Streptococcus, Lactococcus *and* Enterococcus *based on modern classification techniques and serological studies … The typical pathogenic species remained in the genus* Streptococcus *and, with the exception of* Streptococcus thermophilus, *were separated from the nonpathogenic and technically important species of the new genus* Lactococcus *… The 'fecal Streptococci' that were associated with the*

gastrointestinal tract of humans and animals, with some fermented foods and with a range of other habitats, constitute the new genus Enterococcus.

Therefore, when an Irish dairy microbiologist refers to 'fecal Streptococci' he or she *really* means *Enterococcus* spp.! The interest in detecting *Enterococcus* spp. in dairy plants is two-fold: hygiene monitoring; and contamination monitoring of the cheesemaking process. *Enterococcus* spp. display a number of qualities which makes them useful indicators of plant hygiene. They are thermoduric (Boor et al., 2017; Thomas & Prasad, 2014), persist in the processing environment, form biofilms (Flint et al., 2020) and derive from either bovine or human fecal matter (Franz et al., 2003; Maheux et al., 2011). In the context of the Irish dairy industry, where cheddar cheese is the dominant product, their presence in the commercial cheesemaking process is unwelcome, particularly in the finished product where they may cause off-flavors and alterations in texture and appearance (Gelsomino, Vancanneyt, Condon, Swings, & Cogan, 2002). *E. faecium* and *E. faecalis* are also implicated in a number of opportunistic infections of humans and (rarely) foodborne illness (Thomas & Prasad, 2014).

1.3.4.4 Sulfite reducing bacteria/sulfite reducing Clostridia

Sulfite reducing Clostridia (SRCs) are members of the genus, *Clostridium*, which have the ability to reduce sulfite under anaerobic conditions to produce energy (Doyle et al., 2015). Most of the members of the genus which are of interest to the dairy industry are SRCs (Anon, 2014). Because many non-*Clostridium* spp. are able to grow on the media used to select for SRCs, colonies showing positive for sulfite reduction are more correctly referred to as CFUs of "sulfite reducing bacteria (SRBs)/SRCs" (Weenk, van den Brink, Struijk, & Mossel, 1995). Among the species of interest which are detected using the SRB/SRC test are the spoilage organisms *C. butyricum*, *C. tyrobutyricum*, *C. sporogenes*, *C. beijerinikii* and *C. putrifaciens*, as well as the pathogenic species *C. perfringens* and *C. botulinum* (Eisgrubef & Reuter, 1995). The SRB/SRC test is used as an indicator of fecal or soil contamination in dairy plants given that species from the genus, *Clostridium*, are isolated both from the soil and the feces of warm-blooded mammals (Weenk et al., 1995). Interestingly, a survey of sulfite reducing Clostridia (SRC) in New Zealand bulk raw milk concluded that "contamination with SRC is infrequent and at a very low level during milk production processes in New Zealand and these organisms would not be a useful hygiene indicator of process control systems at the farm end of the dairy product supply chain" (Anon, 2014). The question must be asked: what benefit over and above testing for *Clostridium* spp. does testing for SRBs/SRCs confer? The answer is that, historically, it has been more convenient to apply the SRB/SRC plate-based assay than those for *Clostridium* spp. alone, and so detecting SRBs/SRCs became shorthand for detecting *Clostridium* spp. This leads on to the question: will new, rapid methods, especially those based on genetic tests, be able to replicate the broadness of the test for SRBs/SRCs? Or will this exact duplication even be necessary or desirable with the improvements in specificity offered by many rapid techniques (Lavilla et al., 2010)?

1.3.4.5 Yeasts and molds

When writing about yeasts and molds as being of interest to the dairy industry, it is in the context of spoilage of intermediates and finished products rather than as agents of food poisoning (although mycotoxins can be produced by molds; Jakobsen & Narvhus, 1996; Sørhaug, 2011). Yeasts and molds can cause particular problems that bacteria do not pose: they can be highly osmotolerant, tolerate low a_w, temperatures and oxygen tensions, as well as displaying resistance to food preservatives and the ability to grow at low carbohydrate concentrations (Suriyarachchi & Fleet, 1981). They can enter a plant from the air, water, packaging and personnel and may also be difficult to eliminate from a plant once they have got a foothold: they thrive in the moist environments found in dairy plants, living on improperly cleaned and sanitized surfaces (Hernández et al., 2018). A characteristic mold and yeast "house microflora" has been shown to correspond to a particular plant. The main target of molds is cheese, and especially pre-prepared grated cheese (Jakobsen & Narvhus, 1996). The chief genera involved in cheese spoilage are *Penicillium*, *Cladosporium*, and *Phoma*, which mainly attack the product during ripening. Minority genera involved are *Aspergillus*, *Cephalosporium*, *Cladosporium*, *Geotrichum*, *Mucor*, *Scopulariopsis*, and *Syncephalastrum*. It has also been reported that yeast can spoil and, indeed grow undetected in yogurt (Suriyarachchi & Fleet, 1981). Among the genera found growing in yogurt were *Torulopsis*, *Kluyveromyces*, *Saccharomyces*, *Candida*, *Rhodotorula*, *Pichia*, *Debaryomyces*, and *Sporobolomyces*. There is clearly a need to be able to rapidly detect the wide variety of yeasts and molds that trouble the dairy industry, both for plant hygiene and product analysis, especially in light of the mycotoxins some contaminants are capable of producing (Rico-Muñoz, Samson, & Houbraken, 2019). Any rapid method would have to be capable of detection of all of the genera currently capable of growth on the standard media used for testing (Bleve, Rizzotti, Dellaglio, & Torriani, 2003).

1.3.5 Organisms displaying certain growth preferences or resistance as regards temperature

1.3.5.1 Endospores

A number of species form resistant endospores, with their vegetative cells differentiating into multilayered, cryptobiotic forms which are capable of surviving pasteurization, drying, sanitization and many of the other insults directed towards them in the dairy processing environment (Doyle et al., 2015). Under favorable conditions endospores can germinate, outgrow and proliferate, giving rise to spoilage or producing toxins (Thomas & Prasad, 2014). Beyond the best-known genera of *Bacillus*, which are aerobic, and *Clostridium*, which are anaerobic, there are other genera which are of interest to the dairy industry: *Anoxybacillus* (of particular interest is *A. flavithermus*), *Geobacillus* (of particular relevance to the dairy industry is *G. stearothermophilis*) and *Paenibacillus* (Doyle et al., 2015; McHugh et al., 2020). The vegetative cells of sporeformers, once they germinate and begin to grow, display preferences for growth at certain temperatures, leading to the sub-classification of sporeformers into groups based on this (Boor et al., 2017). To add further difficulties, many sporeformers form complex biofilms, where both vegetative cells and endospores are present along with difficult-to-remove glycocalyx (Flint et al., 2020). The principal group of endospores of interest to the dairy industry are the psychrotrophic thermophilic sporeformers (PTS), which are triply problematic in that they are sporeformers (with all the difficulties for the dairy man that this entails) that can survive and multiply (in vegetative cell form) at refrigeration temperatures, e.g. in bulk tank milk, as well as grow at the higher temperatures found in e.g. ultrafiltration plants (Eijlander et al., 2019). Greater than 80% of bulk tank milk has been reported to contain PTS (see Boor et al., 2017). The most commonly isolated members of the PTS are *A. flavithermus* and *B. licheniformis*.

Testing for sporeformers in general and PTS in particular is becoming increasingly important in the dairy industry, especially in HTST processing facilities that effectively control post-pasteurization contamination—it is the carry-through of spoilage and pathogenic sporeformers that are now the major quality and safety issue. The control of pathogens (*B. cereus* and *C. perfringens*) in milk powders (which can regularly contain up to 100 endospores/g, Eijlander et al., 2019), especially those destined for infant formula will also assume greater importance (McHugh et al., 2020).

1.3.5.2 Psychrotrophs

For over 100 years, refrigeration has constituted one of the primary weapons in the dairy industry's war against spoilage and pathogenic microorganisms

(Boor et al., 2017). Holding e.g. bulk tank milk, pasteurized milk products and cheese at refrigeration temperature retards microbial growth, thus prolonging these products' shelf lives. However, there are microorganisms which are capable of growth at these low temperatures (below 7 °C; predominantly Gram negative bacteria) and which constitute a serious risk to consumers' health, as well as causing significant economic losses through spoilage and recalls (Gleeson, O'Connell, & Jordan, 2013; Parente, Ricciardi, & Zotta, 2020). Even though the majority of these psychrotrophs (barring the PTS) are destroyed by pasteurization, their secreted heat-resistant proteinases and lipases and continue to catalyze reactions downstream, causing alterations in product sensory properties (Sørhaug & Stepaniak, 1997). Post-pasteurization contamination with psychrotrophs such *Pseudomonas* spp. occurs through these organisms' persistence in plant equipment, often in the form of biofilms (Boor et al., 2017; Flint et al., 2020). The principal problematic psychrotrophs (outside of the PTS discussed above) are species within the genus, *Pseudomonas*, with *P. fluorescens* the most widely reported spoilage bacterium in raw milk at refrigeration temperatures, other Gram-negative genera (*Achromobacter, Aeromonas, Serratia, Alcaligenes, Chromobacterium* and *Flavobacterium* spp.) and the Gram-Positive bacteria *Corynebacterium, Streptococcus, Lactobacillus* and *Microbacterium* (Parente et al., 2020; Sørhaug & Stepaniak, 1997). Points along the production process of relevance to testing would be raw milk undergoing refrigerated storage, post-pasteurization refrigerated product, and the plant and equipment where biofilms of psychrotrophs would be likely to grow. A single rapid assay which would enumerate and differentiate between the major psychrotrophs encountered in dairy settings would be useful for the purposes of contamination control and monitoring, as would a single convenient, rapid assay for the heat-stable enzymes produced by these bacteria (Wei et al., 2019).

1.3.5.3 Thermodurics
This group of organisms is composed of those which can survive pasteurization and proceed to give problems, either spoilage- or pathogenicity-related, downstream (Gleeson et al., 2013). Sporeformers will be excluded from this discussion, as they have been dealt with above. Thermodurics enter raw milk from the milking parlor environment, principally through contamination of the teat with soil, bedding and feces (Islam et al., 2018; Thomas & Prasad, 2014). Buildups of thermodurics may also occur on milking equipment. It is common to group thermodurics into three

categories: thermophilic thermodurics (optimum growth 50–55 °C; can grow at 40–60 °C); mesophilic thermodurics (optimum growth at 30 °C; can grow at 5–50 °C); psychrotrophic thermodurics (optimum growth 0–25 °C). The non-sporeforming thermodurics are almost exclusively mesophilic, with genera in question being *Corynebacterium*, *Microbacterium*, *Micrococcus*, *Enterococcus*, *Streptococcus* and *Arthrobacter* (Thomas & Prasad, 2014). As well as limiting the shelf life of milk, thermodurics can contaminate post-pasteurization processes and come to reside in process equipment and form resistant biofilms (Flint et al., 2020). Since the majority of thermoduric contamination originates from milking practices, perhaps simple rapid tests could be applied by the farmer at source as a method of self-checking thermoduric contamination.

1.3.5.4 Thermophiles

Thermophiles are those organisms which grow above 40 °C, and which have optimal growth temperatures between 50 and 55 °C (Gleeson et al., 2013). One group of thermophiles—the PTS—have been described above, and so will not be dealt with here; thermophilic thermoduric organisms have been mentioned immediately above. There are references to obligate thermophiles, which have an absolute requirement of growth above 40 °C (Eijlander et al., 2019) and facultative thermophiles, which, as is the case with *Ano. flavithermus* and some strains of *G. stearothermophilus*, may grow at 37 °C (Eijlander et al., 2019). This pair of sporeformers are the most commonly encountered microbes which form biofilms of heated regions (50–70 °C) of milk powder manufacturing plants (Somerton et al., 2012). Many sporeformers are pure thermophiles, i.e. these show no tendency to grow at temperatures below 40 °C, unlike, for example, the mesophiles or PTS (Doyle et al., 2015; Sadiq, Flint, & He, 2018). Non-sporeforming thermophiles include *Ent. durans*, *Ent. faecium*, *Ent.*, *faecalis* and other *Enterococcus* spp., *St. thermophilus* and other *Streptococcus* spp. such as *St. bovis*, *Lb. delbrueckii* and *Lb. helveticus* and *Lysinibacillus fusiformis* (Delgado et al., 2013). Species from genera *Streptococcus*, *Lactobacillus* and *Enterococcus*, as well as being thermophilic are also thermoduric and aciduric—they survive pasteurization and can grow in the acidified environment associated with cheesemaking.

In the last two decades, contamination by thermophilic bacteria of milk powder has become one of the main quality concerns in this area (Flint et al., 2020). Any part of a production process where product intermediate or final product is held above 40 °C for any length of time is vulnerable to the growth of thermophilic bacteria. Thus, the milk powder manufacturing process selects for the growth of these bacteria. Thermophilic bacilli are the predominant spoilage organisms in the final milk powder product, and their presence determines the product selling price (Somerton et al., 2012). Thermophilic bacteria contaminate processes post-pasteurization through two routes: many thermophiles are also thermoduric, even the non-sporeformers (Delgado et al., 2013); and thermophiles can persist in plant and equipment in the form of biofilms (Bassi, Cappa, Gazzola, Orrù, & Cocconcellia, 2017). It has long been recognized that species within the genus, *Streptococcus* and other thermophiles' adhesion to heat exchanger plates in the downstream side of the regenerator section of pasteurizers is responsible for post-pasteurization contamination of milk (Van der Mei, de Vries, & Busscher, 1993). The biofilm-formation propensity of thermophilic sporeformers has also long been recognized (Flint et al., 2020). As with many of the groups of organisms of interest to the dairy industry, the thermophiles are formed by a disparate group of unrelated bacteria, whose only relationship is phenotypic. This adds a layer of difficulty to developing a unified rapid assay for the detection of this group.

1.3.6 General indicators of quality or sanitary status
1.3.6.1 Total plate count
Also known as the spread plate count, total viable count, total bacterial count, aerobic colony count, or aerobic plate count, the total plate count method has been in use in the dairy industry for over a century (Boor et al., 2017; FSAI, 2019; Gleeson et al., 2013). While it has many detractors, the very fact that it is still in use testifies its worth as an assay of the general hygienic status of raw milk, and, on occasion, downstream products. The basis of the test is as an indicator of hygiene and safety in that a high count in e.g. raw milk ($>1.0 \times 10^5$ CFU/mL) suggests poor practices in the milking parlor and/or transportation, whereas a low count ($<5.0 \times 10^4$ CFU/mL) is normally required for a high-quality final product (Gleeson et al., 2013; Willis et al., 2018). It is still in use for the purposes of establishing the hygienic and safety criteria for many national and international regulatory authorities (see Sadiq et al., 2018). The FSAI, for non-fermented dairy products stipulates limits of $<1.0 \times 10^5$ CFU/mL for "satisfactory", $1.0 \times 10^{5-7}$ CFU/mL for "borderline" and $>1.0 \times 10^7$ CFU/mL for "unsatisfactory" (FSAI, 2019). Criticisms of the method includes: anaerobic organisms are not detected, damaged or viable but not culturable (VBNC) microbes may not form colonies on the type of general medium used, only mesophiles are

enumerated, very little information on the types of organisms giving rise to the colonies is provided, is of no value in predicting the presence of pathogens and, in common with all plate-based assays, is labor intensive, time consuming and slow to yield results (Hameed et al., 2018; Sohier et al., 2014; Willis et al., 2018). The total plate count has been one of the first dairy microbiology assays to be replaced by alternative rapid methods, and where these rapid methods have received widespread acceptance. Much bulk tank milk is now tested using FCM to provide the total plate count, and, in addition to this, the 3M Petrifilm aerobic count has in many settings replaced the traditional method (Boor et al., 2017).

1.3.6.2 *Pseudomonas*

Species within the genus, *Pseudomonas*, tick many of the boxes for being a problematic organism for the dairy industry. This genus of spoilage organisms' members are psychrotrophic and notorious for forming biofilms: 89% of *P. fluorescence* isolates from a dairy plant were capable of forming biofilm at 10 °C and 30 °C within 48 h of inoculation (Flint et al., 2020). The biofilms they form tend to be resistant to many commonly used cleaning agents: biofilms of *P. aeruginosa* of dairy origin were resistant to benzalkonium chloride, iodophor and sodium hypochlorite treatment. They are common in raw milk, forming approximately 10% of the microflora, and the majority of Gram-negative organisms, but are killed by pasteurization (Hervert et al., 2016; Sørhaug & Stepaniak, 1997). Their appearance later on in the process is evidence of post-pasteurization recontamination, either as a result of ingress of contaminated material (such as process water) or the spread of biofilm (Kable, Srisengfa, Xue, Coates, & Marco, 2019). *P. fluorescens* is the most widely reported spoilage bacterium in raw milk at refrigeration temperatures and can secrete significant amount of heat-resistant extracellular hydrolytic enzymes such as proteases, lipases and lecithinases in raw milk during storage at low temperature (Sadiq et al., 2018). These enzymes can make their way into pasteurized or UHT milk, and even cause rancidity in frozen butter (Sørhaug & Stepaniak, 1997). Maintaining *Pseudomonas* spp. numbers low in pasteurized milk is the most important factor in prolonging its shelf life, and, thus, it is important to test for this group of organisms in finished product. The utility in testing for *Pseudomonas* spp. is as an indicator for poorly sanitized equipment. The presence of large numbers of *Pseudomonas* post CIP, for example, may indicate that another cleaning cycle is needed (Kable et al., 2019).

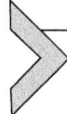

2. Organisms for whom no assay or very few rapid assays exist

2.1 Organisms of interest to the dairy industry for which gaps exist in the assortment of rapid assays

2.1.1 A review of the rapid assays available to dairy industry

Following the reaching of a consensus with the Irish dairy industry as to which organisms/groups of organisms were of interest to them in terms of the routine testing carried out in dairy manufacturing facilities, workers in the Dairy Processing Technology Centre performed an extensive review of the rapid methods commercially available for each organism/group of organisms. As well as performing desk research, manufacturers of rapid microbiological assays were contacted and their involvement in the process sought. Key figures from each of the Dairy Processing Technology Centre's industrial partners were also involved in this work, and their experience in sourcing, validating and implementing rapid assays harnessed. At the end of the process hundreds of rapid assays produced by dozens of companies in a multitude of formats were identified. Fig. 1 shows how many of each of these rapid assays exists for each organism/group of organisms.

As can be seen from Fig. 1, there is a wide disparity between the number of rapid tests available for each organism/group of organisms. Four categories

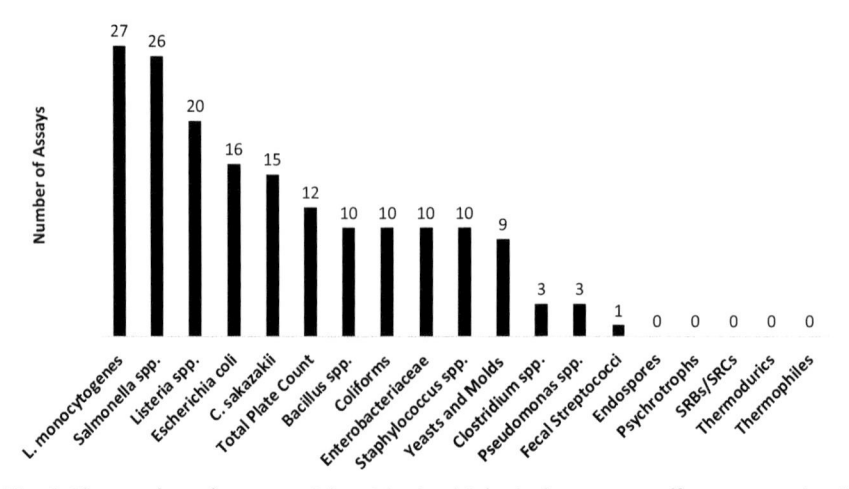

Fig. 1 The number of commercial rapid microbiological assays on offer per organism/group of organisms of interest to the Irish dairy industry. SRBs/SRCs, sulfite-reducing bacteria/sulfite reducing Clostridia.

Table 2 The organisms/groups of organisms of interest to the dairy industry categorized by the number of rapid assays available for their detection.

High coverage (>15 rapid assays)	Medium coverage (5–15 rapid assays)	Low coverage (<5 rapid assays)	No coverage (0 rapid assays)
Escherichia coli	*Bacillus* spp.	*Clostridium* spp.	SRB/SRCs
Listeria spp.	Coliforms	Fecal streptococci	Endospores;
Listeria monocytogenes	*Cronobacter sakazakii*	*Pseudomonas* spp.	Psychrotrophs
Salmonella spp.	Enterobacteriaceae		Thermodurics
	Staphylococcus spp.		Thermophiles
	Total plate count		
	Yeasts and molds		

SRB/SRCs, sulfite-reducing bacteria/sulfite-reducing Clostridia.

could be designated to summarize the availability of rapid assays per organism/group of organisms: high coverage (>15 assays available); medium coverage (5–15 assays available); low coverage (<5 assays available); no coverage (0 assays available). The organisms/groups or organisms thus arranged are shown in Table 2.

Certain patterns are clear from the data. "High coverage" organisms/groups of organisms are all pathogens for which strict regulations exist regarding their levels in finished product. Three out of six "medium coverage" organisms/groups of organisms are also regulated-for pathogens, while there exist regulatory limits for the Enterobacteriaceae and total plate count. Among the "low coverage" group, only one regulated-for pathogenic genus (*Clostridium*) is found out of the three organisms/groups of organisms. Again, only *Clostridium*, which is found within the SRB/SRCs constitutes a pathogen out of the "no coverage" group, which contains five organisms/groups of organisms. Therefore, species or genera among which are found pathogens the presence of which is regulated-for tend to have a selection of commercial rapid assays available for their detection.

In the "medium coverage" category are also found three broad taxonomic groupings—coliforms, Enterobacteriaceae and yeasts and molds—along with a general indicator of quality or sanitary status—the total plate count. The fact that so many commercial assays exist for these non-pathogens indicates the importance to the dairy industry of monitoring for these groups of organisms. The total plate count, coliform count and Enterobacteriaceae count are extensively used as indicators of raw material quality, plant hygiene, product intermediate quality and product shelf life. Lapses in any of these give rise to economic losses and so it has clearly been of interest to the dairy industry to streamline testing for these. Very obviously developers and manufacturers of rapid microbiological assays have recognized this need and responded to it. Similarly, the presence of yeasts and molds in a dairy manufacturing facility are highly unwelcome and problematic to eliminate once established (see above). The importance of rapid monitoring for these is reflected in the relatively wide range of commercial rapid assays for their detection.

The question then arises: why have commercial developers and manufacturers of rapid microbiological assays under-provided for in terms of the organisms/groups of organisms found in the "low coverage" group and ignored the "no coverage" group?

2.2 Organisms/groups of organisms with low coverage of rapid assays

2.2.1 Clostridium *spp.*

There are a number of motivations for testing for species within the genus *Clostridium*.

Regulations specify the levels of *C. perfringens* in ready-to-eat food (FSAI, 2019). Producers of dairy powder are interested from their own quality systems perspective in knowing the numbers of members of this genus of sporeformers making their way into final product. Additionally, large customers of dairy powder producers such as infant formula manufacturers may stipulate maximum levels of Clostridia for a supplied product. Since *Clostridium* spp. find their way into milk from contamination of the teat (Reineke & Mathys, 2020), their presence can be used as an indication of lapsed hygiene at milking. This current study found that there were three rapid commercial assays for the detection of *Clostridium*: Biotekon's Foodproof® Kit (PCR-based); CHROMagar™ *C. perfringens* rapid plate-based assay; Sylab's BacTrac 4300 Microbiological Impedance Analyzer-based system. As described above, the detection of *Clostridium* is not trivial. Dairy manufacturers are also in disagreement over whether it is useful to test for *C. perfringens* only, diverse members of the genus, *Clostridium*, or the SRCs (Doyle et al., 2015). Perhaps this uncertainty, combined with the intricacies of clostridial taxonomy is responsible for the small number of rapid methods for the detection of *Clostridium*.

2.2.2 Fecal streptococci

It is one of the quirks of the Irish dairy industry that fecal Streptococci (a synonym for species within the genus, *Enterococcus*) are one of the groups of organisms regularly tested for, especially in a cheesemaking context. Testing is concerned with hygiene monitoring and detection in finished product. According to the research carried out for this study, one commercial rapid method exists for the detection of fecal Streptococci: the bioMérieux Vitek® 2 Compact platform, an automated system based on the identification of organisms' enzyme repertoire, allows the identification of species within the genus, *Enterococcus*. Why more rapid methods for the detection of fecal Streptococci have not been commercialized is possibly down to the reduced size of the market. Were dairy producers outside of Ireland to adopt testing for *Enterococcus* spp. then there would be more incentive for other companies to develop rapid assays for members of this genus. Technical difficulty would not be a reason for the paltry offering of rapid techniques for *Enterococcus*: biochemical/enzymatic/phenotypic, PCR-based, or antibody-based assays to detect members of a genus are already commercialized for other genera such as *Listeria*.

2.2.3 Pseudomonas *spp.*

Testing for species within the genus, *Pseudomonas*, is for both quality/shelf-life and hygiene monitoring. It is surprising that there only exist three rapid assays for this genus. These are: CHROMagar™ *Pseudomonas* spp. rapid plate-based assay; Sylab's BacTrac 4300 Microbiological Impedance Analyzer-based system; Sylab's RiboFlow® rRNA-detecting lateral flow assay. In principle, rapid testing for species within the genus, *Pseudomonas*, should not be laden with the difficulties reported for species within the genus, *Clostridium*, as the same controversies with taxonomic designation of strains has not been reported (Sørhaug & Stepaniak, 1997). It could be concluded that either the dairy industry's demand for rapid *Pseudomonas* assays has not been picked up on by kit manufacturers or that the dairy industry is content with the current selection of assays.

2.3 Organisms/groups of organisms with no coverage of rapid assays

2.3.1 Endospores

It is most likely due to the technical difficulty of detecting all (or the majority of) endospores in a sample that no rapid method exists for their detection. There are two areas of difficulty in detecting endospores: the fact that endospores constitute one of the two forms in which a strain exists; and when we speak of detecting all of the endospores in a sample we are talking of detecting bodies from several genera with widely differing growth requirements, metabolisms, morphologies, surface antigens, sizes etc. (Eijlander et al., 2019). A standard PCR will not discriminate between DNA isolated from a vegetative cell from that isolated from an endospore. Unless a method such as density-gradient centrifugation is used to separate the vegetative cells from endospores in a sample, PCR as a rapid method is of no use in endospore detection. Similarly, growth-based rapid assays must remove or inactivate (through heating) vegetative cells from a sample in order to yield data on the growth of only endospores (Thomas & Prasad, 2014). However, since endospores have widely differing growth requirements (species within the genus, *Clostridium*, are strictly anaerobic, while *Bacillus*, are aerobic) it is very difficult to envisage one single growth assay successfully detecting all endospores present—unless this were to come in a multi-well format where the medium, temperature and atmosphere of each well was tailored to the growth of a different strain. Any immunoassay would require a cocktail of antibodies of broad (genus) specificities in order to detect the common species from the common endospore genera (*Anoxybacillus, Bacillus, Clostridium, Geobacillus, Paenibacillus*) as it would be very unlikely to find one antibody of acceptable specificity which bound to all endospores (Kennedy & Wilkinson, 2017). Interestingly, were the problem of separating endospores and vegetative cells overcome, PCR could be used to detect a broad range of endospores: a cocktail of primers could be applied to amplify the DNA of multiple strains. Were technical issues regarding the staining of endospores overcome, the technique made more convenient and sample reading automated, fluorescent in-situ hybridization (FISH) could be used to detect all endospores: probes are available for broad taxonomic groups such as that into which all endospore formers can be placed, the phylum Firmicutes (Doyle et al., 2015; Rohde, Hammerl, Appel, Dieckmann, & Al Dahouk, 2015).

2.3.2 Psychrotrophs
This group is composed of a wide range of unrelated bacteria, whose only shared feature is the ability to grow at temperatures below 7 °C (see above). So broad is this group taxonomically (both Gram-positive and -negative bacteria are psychrotrophs) that any rapid assay must surely be growth based. The difficulty in developing a *rapid* assay where growth temperatures must be maintained below 7 °C is the slow rate at which growth would be

detectable. This would suggest that rapid assays for psychrotrophs could be based on highly sensitive electrochemical, spectroscopic or colorimetric measurements to detect changes in the growth medium caused by these organisms' metabolism or a role for novel biosensors (Ziyaina et al., 2020).

2.3.3 SRB/SRCs

This group of organisms is characterized by two traits of metabolism—the ability to reduce sulfite and do so under anaerobic conditions (see above). The dairy industry has been taking advantage of these traits, using solid media containing sulfite, incubated in an anaerobic environment, to detect the members of the genus, *Clostridium*, of most interest to them. Unfortunately, a number of non-*Clostridium* species can also grow on such media and so an element of SRB/SRC counts includes these non-Clostridia (Doyle et al., 2015). The fact that no rapid assay exists for SRB/SRCs probably reflects the fact that very few alternative methods could hope to replicate an assay based on the appearance of black colonies on solid medium. Such rapid assays that could mirror the current standard assay would necessarily be growth based, and as suggested above for the psychrophiles, these would need to be based on a highly sensitive method of detecting metabolism. PCR or antibody-based assays could theoretically form the basis of an assay for SRCs only, which would be more informative for the dairy industry, as non-clostridial SRBs would be omitted from the count.

2.3.4 Thermodurics

Thermodurics constitute another broad church of unrelated microorganisms whose only common traits are their presence in raw milk and their ability to survive pasteurization (see above). Any rapid detection method would necessarily involve sample pasteurization followed by the implementation of detection technology. Similar to the psychrotrophs, neither molecular nor immunological techniques would likely achieve success in specifically detecting these organisms. Any method, however, which could immediately detect and quantify the presence of live cells and intact endospores post heating holds the possibility of rapid detection. Such methods would include FCM combined with viability dyes (Kennedy & Wilkinson, 2017), fluorescence microplate-based or microfluidics-based methods using similar viability dyes, or again, sensitive methods of growth detection. One issue with any growth-based method, and this holds for the detection of other broad groups, would be settling on a growth medium suitable for the survival, growth and reproduction of a diverse range of strains, spread across distantly related genera (Hervert et al., 2016). This would not be a trivial and uncontroversial matter.

2.3.5 Thermophiles

The detection of this group presents similar problems as the detection of the psychrotrophs. As with the latter group, rapid detection methods would have to be growth-based. A possible detection method for both groups would be to allow a sample to grow above 50–55 °C for a short number of h and then detect any viable cells using FCM, an automated imaging technique, microfluidics or a sensitive spectrophotometric or fluorometric plate-based method. As with the thermodurics, the choice of growth medium for such an assay would be crucial.

3. Alternative microbial assay validation

3.1 Regulations surrounding alternative microbial test methods

European Regulation (EC) No 2073/2005 on microbiological criteria for foodstuffs sets out requirements for dairy companies and food businesses in general for multiple aspects associated with the provision of microbiologically safe food to consumers, including requirements surrounding the analytical methods that can be used when testing foods (European Commission, 2005). Reference test methods are provided by Regulation No 2073/2005 for microbiological tests for individual species or groups of species in broad classes of foods, the most recent version of which should be used to test for the presence of that particular microbe. However, alternative test methods may be preferred by the FBO, as they may allow a shorter TTR or higher throughput of test samples than the recommended reference test method. According to the regulation, alternative methods to the reference method can be used as long as they provide at least equivalent results to the reference method in the relevant food category, as demonstrated by validation against the most recent edition of the analytical reference method specified in the regulation (European Commission, 2005).

3.2 Paths for adoption of rapid test methods

European and international third-party accreditation and certification organizations such as AFNOR, Association of Analytical Chemists (AOAC), MicroVal and NordVal can carry out independent unbiased validation of proprietary alternative methods according to ISO 16140:2016 or similar standards in line with the abovementioned regulation (ISO, 2016). The performance of these validation procedures results in issuing of an independent certificate of method performance, confirming that the method meets an appropriate standard for its intended use, as set out in the criteria of the validation study. FBOs and independent testing laboratories can then perform method verification studies to confirm that the validated method functions in the end users' lab as was determined in the validation study. In the case where a company wishes to use a method that does not carry these third-party certifications, validation using the ISO 16140 series of alternative method validation standard procedures, or equivalent recognized protocols, must be carried out and the method must be authorized by the relevant authority overseeing compliance with Regulation No 2073/2005 (FSAI, 2014).

In certain instances, alternative microbiological methods can present challenges to alignment with strict regulatory terminology. In many cases, alternative microbiological methods can present microbiological data in forms different to that produced by the standard culture-based methods (i.e. CFU per sample). For example, ATP-based bioluminescence test systems produce results in Relative Light Units (RLU; Bottari & Santarelli, 2015), and FCM-based BactoScan™ and BactoCount instruments present results of bacterial counts as Individual Bacterial Counts (IBC). The microbiological criteria for food groupings set down in Regulation (EC) No 2073/2005 specify microbial limits in CFU per sample. Similarly, the microbiological criteria for raw milk hygiene set down in Regulation (EC) No. 853/2004 (European Commission, 2004) indicate the maximum number of microbes that are allowable based on a plate count per ml at 30° C—therefore in units of CFU/mL. Dairies commonly employ a payment scheme for farmers based on the total bacterial counts (CFU/mL) in raw milk. However, rapid high-throughput alternative FCM-based BactoScan and BactoCount instruments are widely used in dairies and testing laboratories for microbial counting in raw intake milk, where results are presented as IBC/mL, the values of which can differ from CFU/mL. Conversion factors have been developed to equate IBC to CFU, and a standard method (ISO 21187; ISO, 2004) is available to allow determination of conversion factors between standard method results and alternative method results in line with regulations. However, variability can still exist in conversion factors between different labs that have individually validated their conversion factors according to the regulations using the recommended standardized methods. This has led to a drive to harmonize conversion factors at least at a national level (Madden, Gordon, & Corcionivoschi, 2017).

In recent years, a standard FCM method for enumeration of lactic acid bacteria in starter cultures, probiotics and fermented dairy products has been developed (ISO 19344; ISO, 2015), which presents microbial counts in active fluorescent units (AFU) and total fluorescence units (TFU). Recommendations for probiotic product labelling include reference to microbial content in terms of mass or CFU. Furthermore, clinical data on the efficacy of specific amounts of probiotic preparations is mostly provided in CFUs, so for comparison purposes, data on microbial load in products should be relatable to this (Jackson et al., 2019). It has been suggested that AFU and CFU values may correlate well for fresh microbial preparations, but with processing, and increasing storage and shelf life, the relationship is not equivalent, with the emergence of viable but non-culturable cells, detected by FCM as active cells, but not detected by plate-based methods as CFU (Jackson et al., 2019; Wilkinson, 2018). Therefore, further work on the harmonization of FCM and plate counting data needs to be carried out, or the acceptance of FCM data by further demonstration of its virtues independent of correlative studies.

Coordination of product end-users, product developers, standards-developing bodies and third-party validation bodies with the relevant authorities to determine what is required to allow a specific method offering alternative results to CFU/mL to be accepted legally as a replacement to standard methods for product release would be beneficial before lengthy validation and verification procedures are carried out on alternative test systems. Furthermore, the implementation of rapid test methods in the dairy industry is not only dependent on their adherence to regulatory requirements as determined for broad food groups according to the regulations, but at a practical level, their compatibility with individual assay criteria, such as specific dairy matrices, determines their suitability to routine testing regimes to produce results in line with regulations.

4. Effects of dairy matrices on assay performance
4.1 Dairy products for microbiological testing

The range of products offered by dairy companies include those for direct consumption by end users, such as heat-treated milk, cheese varieties and butter, but also food ingredients such as milk and protein powders, concentrates and isolates that are used for manufacture of, for example, nutritional supplements, infant formula and baked goods. A range of dairy intermediates, ingredients and products were identified by the consortium of dairy companies which form the DPTC's industrial partners that are regularly tested using microbiological methods, and so should, from their perspective, be included as test matrices in validation studies of alternative microbial methods, to demonstrate their level of compatibility with the technology (Fig. 2). These included raw and pasteurized milk, cheese and butter, as well as dairy powders such as skimmed milk powder, concentrates of milk proteins, and isolates of whey, lactose and casein. From a review of commercially available alternative microbiological technologies, the number of rapid alternative test systems that have undergone validation studies using each of these matrices is shown in Fig. 2. While matrices such as raw milk, cheese and pasteurized milk are extensively used in validation studies of dairy, food ingredients such as enriched milk powder (EMP; also referred to as fat-filled milk powder), milk protein concentrates (MPC), whey protein isolates (WPI) and skimmed milk concentrates (SMC) are included less often.

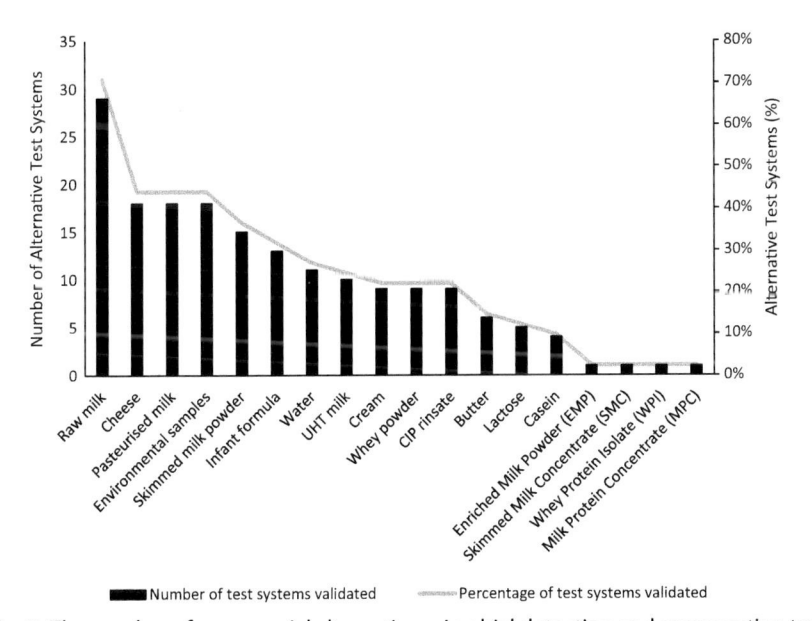

Fig. 2 The number of commercial alternative microbial detection and enumeration test systems that have performed and published validation studies using the dairy matrices of interest to industry members of an Irish dairy consortium (DPTC). The percentage of the total test systems investigated in the current study that performed validation studies with each of the matrices is also shown needed (Kable et al., 2019).

For most alternative testing platforms, introduction of a dairy matrix as a test sample results in an increase in microbial detection limits compared to pure culture alone, and this is illustrated in the literature. In the case of isothermal amplification methods, the detection limit for *L. monocytogenes* in a milk powder matrix using a propidium monoazide loop mediated isothermal amplification (PMA-LAMP) method was 10-fold higher than in a broth culture (Wan et al., 2012). Detection limits for a range of bacteria using LAMP were similarly found to be higher in skimmed milk, whole milk, and a range of cheeses, and *L. monocytogenes* enriched to a level of 4–5 CFU/mL was undetectable above a background level in milk and cheeses of varying fat content including mozzarella, crescenza and cottage cheese by isothermal amplification, while this concentration of bacteria was detectable in a broth culture (Tirloni et al., 2017). Studies of endpoint PCR incorporating immunomagnetic separation (IMS) demonstrated a 10-fold higher detection limit compared to pure culture for *L. monocytogenes* in milk (Luo et al., 2017), and for *B. cereus* in pasteurized milk (Forghani et al., 2015). The sensitivity of a nanozyme lateral flow assay for *E. coli* O157:H7 was 0.95×10^2 CFU/mL in buffer, and 9×10^2 CFU/mL in milk (Han et al., 2018). However, there are also examples of test systems demonstrating similar performance in a dairy matrix as in pure culture. The detection limit for emetic *B. cereus* in milk was found to be the same as in broth for a PMA-qPCR assay (Zhou et al., 2019), and a lateral flow immunoassay for *L. monocytogenes* had a similar detection limit in 2% reduced fat milk as in buffer (Cho & Irudayaraj, 2013). What is clear is that there is variability in microbial detection levels dependent on the sample matrix, the specific detection system, the procedures in use and the target microbe, and validation of a test system in specific matrices for specific microbes is required before use of the test system for routine analysis in that matrix can be implemented.

4.2 The effect of dairy product varieties on alternative test system performance

The physical and biochemical characteristics, as well as the composition of different dairy matrices can affect the performance of rapid microbiological technologies. Depending on the specific product, dairy products vary in pH, a_w, and in fat, lactose, casein and whey compositions, and even products generally referred to as MPC can comprise items ranging from 40% to 90% protein content. Furthermore, the composition and structure of milk constituents, such as casein micelles and soluble casein proteins are affected by calcium ion concentrations, by temperature, and by the pH of the dairy product (Lin, Leong, Dewan, Bloomfield, & Morr, 1972; Marchin, Putaux, Pignon, & Léonil, 2009; Walstra, 1990). These differing properties can affect the microbial detection and enumeration efficiency of test systems in specific dairy matrices, and the differentiation of microbial cells from matrix components.

At a basic level, the source of the milk may define the performance of a test system—it has been suggested that the performance of FCM-based systems, such as the BactoScan™, may vary if, for example, the test sample is sheep's milk instead of cow's milk (Tomáška et al., 2006). The physical structure of the dairy matrix can also affect test system performance. The viscosity of undiluted milk and dairy products such as ice cream and yogurt has been shown to negatively affect the smooth flow of analyte on lateral flow strips for *S. aureus* enterotoxin A (Upadhyay & Nara, 2018), the velocity of the reagents across the membrane for detection of *E. coli* O157:H7 in milk (Xue, Zhang, He, Wang, & Chen, 2016) and the ability of the analyte to reach the detection pane in a lateral flow assay for *S. aureus* enterotoxin B (Chiao, Wey, Tsui, Lin, & Shyu, 2013).

The presence or absence of product treatments, such as heat treatments, can affect the ability of test systems to produce accurate results. In a comparative study of rapid total bacterial enumeration systems versus the standard plate counting method, the MPN-based TEMPO system proved to correlate better with the standard plate count for heat-treated milks than the BactoScan™ FCM-based system, which correlated with the standard culture technique on data from raw milk samples only (Loss, Apprich, Kneifel, Von Mutius, & Genuneit, 2012). Conversely, from a study of a commercial ELISA kit for detection of *S. aureus* enterotoxins in dairy products, it was suggested that endogenous alkaline phosphatases and lactoperoxidases in raw milk and cheeses made from raw milk may interfere with the performance of ELISA assays (Hennekinne et al., 2007). Further properties such as pH may also affect test system performance. In an enrichment ELISA for *S. typhi*, a microbial level of 1.0×10^2 CFU/mL could be detected in raw milk but not in curd following a six-h enrichment, possibly due to the low pH of the curd matrix, or the presence of high numbers of other bacteria such as Lactobacilli (Kumar, Balakrishna, & Batra, 2008).

Immunological technologies, such as ELISA, ELFA and FCM, and nucleic acid-based technologies, such as PCR and isothermal amplification, have been shown to be sensitive to fat and protein components in the dairy matrix (Gunasekera et al., 2000; Paul, Van Hekken, & Brewster, 2013; Soejima, Minami, & Iwatsuki, 2012), preventing discrimination of cells from the matrix, disrupting the proper functioning of the detection method itself, or producing false-positive results due to interactions of matrix components with detection reagents such as antibodies (McClelland et al., 1994).

Matrix interference effects on test system efficiency may also be dependent on the target microbial species. In a study of an indirect impedance system, *L. monocytogenes* and *L. plantarum* were not detectable in UHT milk by the system, despite reaching 8–9 log CFU/g (Johnson et al., 2014), while for other test microbes including strains of *B. cereus* and *E. coli*, the time to detection by the system was similar in UHT milk and in pure culture. The authors suggested that this might be due to growth conditions affecting microbial CO_2 production. They highlighted the importance of prescreening test systems with the microbe and matrix of interest before selecting the system for routine use (Johnson et al., 2014).

The physico-chemical properties of individual dairy matrices can also affect efficiency of culturing steps used in traditional and alternative test systems. Demonstrations of test system detection limits incorporating a culturing step should be carried out, including specific food matrices, rather than just using pure culture. The importance of this was demonstrated in a study by Nyhan and colleagues (2018), who showed that the growth of a cocktail of *L. monocytogenes* and *L. innocua* in Béarnaise sauce and zucchini paste was consistently lower than that in BHI medium across a range of pH, a_w and undissociated acid manipulations, indicating that food matrix structure and composition affects the rate of microbial growth (Nyhan et al., 2018). Furthermore, in a study by Tirloni et al. (2017), differences in enrichment levels of *L. monocytogenes* were noted in a range of milk and cheese matrices, with cheeses such as Taleggio and Gorgonzola showing up to 4-log lower enrichment levels of the microbe in Fraser broth at 48 h compared to whole milk, ricotta and mascarpone (Tirloni et al., 2017). This not only has implications for traditional culture-based detection systems, but alternative platforms and test systems relying on an enrichment step will be affected by the ability of target bacteria to grow at a reasonable rate in the conditions determined by the specific properties of the dairy matrix.

The variability in microbial test system performance based on specific matrix properties, together with the level of compositional variety in dairy products on offer justifies including as many dairy matrices as possible in method validation studies to ensure their compatibility with the alternative test systems. Conversely, the breadth of the product range presents method development companies with problems related to the time, money and personnel that would be required to perform such extensive studies per organism, also being cognizant of the fact that "dairy" as a group of products may be just one of many food groups under examination in validation studies. Inevitably, the value of validating methods in specific matrices is influenced by the value of that product to the majority of the test system manufacturer's customer base. Specific food and dairy product popularity and production levels can vary geographically based on cultural preferences, lifestyle choices and economic factors, and while certain products may represent a major product regionally, on a global scale they may only constitute a minor proportion of food products. However, based on the variability in system and assay performance dependent on matrix properties among other factors, prior verification of the compatibility of validated rapid test system and alternative assays with matrices of interest is of vital importance to dairy industries, before an alternative test system is adopted.

4.3 Strategies to reduce dairy matrix interference

To minimize the interference from the dairy matrix on the performance of microbial detection systems, methods for removal, reduction, inactivation or suppression of interfering components of the dairy matrix are often included in sample preparation. The type of method and the scale of pretreatment is based not only on the complexity of the dairy matrix, but also on the concentration and type of the target microorganism, whether detection or enumeration of the microbe is required, or if absence of the microbe has to be demonstrated. For a commercial test system to be applied to a comprehensive collection of microbial tests routinely carried out within the dairy industry, a range of sample treatment methods may have to be incorporated that are specific to individual tests, and have been demonstrated to function effectively in individual specific dairy matrices.

In cases where the target organism is present in large quantities, such as starter culture preparations, fermented dairy products and probiotic preparations, dilution of the matrix can be performed which reduces background signal due to the matrix, while still providing sufficient microbial levels for detection and enumeration (Casani, Flemming Hansen, & Chartier, 2015; Geng, Chiron, & Combrisson, 2014; Wilkinson, 2018). Dilution has also been reported to effectively promote the smooth movement of viscous samples such as milk, yogurt and ice cream across lateral flow assay strips while also allowing detection of *S. aureus* enterotoxins (Boyle, Njoroge, Jones, & Principato, 2010; Jin et al., 2013; Upadhyay & Nara, 2018). However, even with dilution, certain dairy matrices may still pose problems for specific assay accuracy and precision. In an interlaboratory study of an FCM method for enumeration of lactic acid bacteria in starter cultures and fermented products, difficulties were encountered in discriminating total bacterial cells from yogurt particulate material even after dilution of the samples, resulting in higher repeatability and reproducibility values for total fluorescence units, but not active fluorescence units in yogurt samples compared to frozen or freeze-dried starter culture preparations (Casani et al., 2015). In other cases, bacteria can be present in such high concentrations as to overcome any possible matrix interference without pre-treatment, as demonstrated by the detection of *S. aureus* at a concentration of 1.0×10^5 CFU/mL in UHT milk by a qPCR method without any sample pre-treatment (Dong et al., 2018). Mass spectrometry (MS) instruments and technologies for food microbial analysis are primarily targeted at rapid food isolate identification to the genus, species and even sub-species level, and confirmation of preliminary detection assays, as an alternative method to biochemical microbial identification and confirmation methods (Jadhav et al., 2015). MS has been used to identify bacterial species isolated from probiotic drinks and yogurts (Angelakis, Million, Henry, & Raoult, 2011), and for the rapid identification of mastitis-causing pathogenic isolates from milk (Barreiro et al., 2010). In these cases, the method itself, and its application in dairy microbiology is not directly subject to dairy matrix interference, as the matrix does not come into contact with the MS system. The dairy matrix may affect the culturing of specific bacterial species, as outlined earlier, for generation of isolates for subsequent MS analysis.

Where the target microorganism is present in the dairy matrix in lower numbers that preclude dilution of the matrix, dairy matrix components exert a stronger effect on the capabilities, and functioning of the detection systems; they can also mask the target organism, resulting in false negative results, or themselves be erroneously detected as the target organism, and so produce false positive results. As discussed earlier, for most microbiological assays, detection limits for microorganisms in dairy matrices are usually higher than for broth cultures.

When absence of a particular microbe in a volume of the dairy sample has to be confirmed, broth enrichment procedures are usually included in the detection protocol to increase the target microbe to levels that are detectable by the assay technology. Enrichment protocols are used in traditional culture-based methods, but are also incorporated into alternative microbial test platforms including PCR, isothermal amplification, ELISA, FCM and lateral flow assays to ensure detection of any target cells that may be present in the sample (El-sharoud, 2015; Liu, Sui, Wang, & Gu, 2019; Shan et al., 2016; Song et al., 2016; Tirloni et al., 2017; Yu et al., 2017; Zhang et al., 2016). Enrichment can last from 2 to 48 h, depending on the microbe of interest and the detection limits of the assay technology. While this step increases the TTR of the assay, it results in dilution of the food matrix with culture broth, and an increase in target cell numbers, which can aid downstream detection by the assay technology. A study of a qPCR-based *Salmonella* detection method showed that an assay incorporating a 4 h broth enrichment step allowed detection of the pathogen in milk matrices of varying fat (2–6%) and solid-not-fat (9–11%) content, at levels of 1 CFU 25/g irrespective of the milk matrix (El-sharoud, 2015). Similarly, matrix interference from milk proteins was reduced in a proof-of-concept study of direct MS microbial analysis of enrichment cultures of milk, only after performance of a secondary enrichment step to further dilute the interfering particles in the milk matrix, which masked identification of species-specific peptide mass fingerprints (Jadhav, Shah, Karpe, & Morrison, 2018).

If enumeration is required and, thus, enrichment cannot be carried out, or where extra matrix cleanup measures in addition to dilution by enrichment are required, physical and chemical matrix clearance methods have been tested for removal and reduction of particulate material, fat and protein. For many platforms such as PCR, isothermal amplification, ELISA, FCM and lateral flow, strategies to remove these components include filtration to remove large particulate material, centrifugation to separate and remove the fat component, treatment with detergents such as SDS or Triton X-100 to solubilize lipids, treatments with EDTA or citrate to solubilize casein micelles, and treatment with proteases such as proteinase K or savinase to digest the protein component (Bosward, House, Deveridge, Mathews, & Sheehy, 2016; Chiao et al., 2013; Gunasekera et al., 2000; Kumar & Kumar Mondal, 2015; Liang et al., 2015; Liu et al., 2019; Paul et al., 2013; Soejima et al., 2012). To use bioluminescence-based assays in milk samples, pretreatment of the milk may require the lysing of somatic cells and degradation of the somatic cell ATP, prior to extraction of the bacterial ATP for measurement of microbial contamination without interference from somatic cells. Centrifugation and density gradient centrifugation to isolate bacterial cells may also be carried out (reviewed in Bottari & Santarelli, 2015). Freezing has also been used as a method to "defat" the milk, by removal of the crystalized fat globules from the surface of the thawed milk for analysis using a lab-on-a-chip FCM test system (Fernandes et al., 2014). Immunomagnetic separation of target cells of interest can also be employed to extract and concentrate the microbial cells from the complex matrix, and has been demonstrated coupled to FCM (Seo, Brackett, & Frank, 1998), PCR (Luo et al., 2017) and ATP measurement test systems (Bottari & Santarelli, 2015). Treatment with HCl to reduce the pH and precipitate casein has also been used for ELFA methods (Hennekinne et al., 2007).

Pretreatment strategies can use a combination of these methodologies to reduce background interference for subsequent microbial assays. The individual matrix component proportions in different dairy products, and the presence of supplemental ingredients, such as vegetable fats and vitamins in EMP, could affect the capability of technologies to assess the microbial content, and the strategies employed for matrix preclearance. In addition to this, conformational changes in matrix components, such as proteins, in response to heat treatments can alter the effect these pre-clearance mechanisms have on the effectiveness of the preclearance protocols. In a study by McClelland et al. (1994), a milk-clearing reagent was effective in removal of interfering particles from pasteurized milk for detection by FCM, but not from ultra-heat treated milk, where the clearing solution had no effect on removal of interfering particles, and produced results similar to an un-cleared control. The authors attribute this to the differential behavior of casein micelles following heat treatment that may have prevented their flocculation and removal. Similarly, homogenization alters the structure of casein micelles, which may also affect the ability to remove these particles. Care should be taken that the agents used for removal of interfering components, do not themselves interfere with the proper functioning of the detection assay, or result in excessive microbial cell loss, for example loss of cells present in the discarded supernatant of a centrifuged sample, or loss of cells attached to matrix components that are removed from the sample (Bosward et al., 2016; Geng et al., 2014). Even in cases where pretreatments are performed, detection limits in the pretreated matrices can still be higher than those in pure culture. In a qPCR assay for detection of *Enterobacteriaceae*, milk was treated with proteinase K following centrifugation to remove micellar casein, and sample pellets decreased in size due to degradation of matrix protein. Even with this treatment, the Ct values of the qPCR assay were higher in the milk matrix than in sterile water (Soejima et al., 2012). However, reduction of detection limits in dairy matrices to levels comparable to pure culture may not be a goal; it may be sufficient to include matrix pretreatment steps to reduce matrix interference enough to allow the alternative test system to have an advantage over traditional methods in aspects such as time-to-result, sample throughput and test system footprint.

5. Conclusions and recommendations

Eight of the 19 organism/groups of organisms of interest to the Irish dairy industry have between zero and five commercially available rapid assays for their detection. Even though a number of problematic organisms/groups of organisms belong to these under-provided-for groupings (e.g. *Pseudomonas* spp., endospores, thermodurics, *Clostridium* spp.) commercial kit, instrument and platform manufacturers have been slow in developing rapid alternative methods for them. While it may be technically difficult to develop rapid assays for broad, ill-defined groupings of organisms, we would urge commercial assay developers to take note of the evidence provided by this study and develop assays to fill the evident gap, especially given the increasing focus on a number of these grouping. For example, testing for sporeformers is becoming increasingly important in the dairy industry, where their carry-through post-pasteurization into powders is of growing concern, and, in a context where the microbiological quality of raw milk is higher than it has ever been, the focus is shifting to spoilage organisms which enter the product post-pasteurization. There is also a growing clamor for rapid methods to detect fecal streptococci.

Culture-based methods of microbial analysis have been in use for over 100 years and are widely accepted as the gold standard of microbial detection and enumeration. Microbiological criteria in official regulations concerning safe levels of microbes for consumption and microbial limits for sale of dairy products are founded on culture-based test methods. Novel technologies that produce results that do not correlate exactly with this gold standard, and produce results based on measurement of other cellular characteristics, such as cell membrane permeability, intracellular enzyme activity, or ATP concentration can encounter obstacles to implementation in routine testing regimes, due to non-adherence to strictly defined regulation criteria. Specific terminology in official regulations, such as explicitly requested units of measurement of microbes, may create difficulties in the acceptance by authorities of alternative methods that do not report data in CFU by authorities, and therefore by laboratories testing dairy products. Furthermore, a lack of demonstrated assay performance in a diverse range of dairy products routinely tested by the dairy industry can prevent the industry from investing time and money in a test system that may not prove compatible with their range of products when in-house verification studies are performed. Therefore, further evidence of assay performance in a broader range of dairy matrices could help the industry to make the change to alternative microbiological testing systems.

It is also key that the dairy industry customer base is open to the adoption of rapid methods to replace the current traditional ISO cultural methods prescribed in product specifications. It would also be beneficial for the ISO working groups (e.g. those of the International Dairy Federation) to begin to evaluate and propose rapid microbial methods on the ISO agenda for adoption as standard test methods. Finally, a forum wherein the dairy industry could engage with commercial test kit providers and inform them of their specific requirements regarding certified rapid assays would represent a significant first step in manufacturers offering a wide choice of rapid methods to this important sector.

References

Angelakis, E., Million, M., Henry, M., & Raoult, D. (2011). Rapid and accurate bacterial identification in probiotics and yoghurts by MALDI-TOF mass spectrometry. *Journal of Food Science*, 76(8), 568–572.

Anon. (2014). *Survey of sulfite reducing Clostridia (SRC) in New Zealand bulk raw milk: MPI technical paper no: 2014/40; Prepared by Food Risk Assessment group*. Wellington: Publications Logistics Officer, Ministry for Primary Industries.

Anon. (2019). *Guidelines for the validation of analytical methods for the detection of microbial pathogens in foods and feeds; edition 3.0*. U.S. Food and Drug Administration Foods Program.

Artursson, K., Schelin, J., Lambertz, S. T., Hansson, I., & Engvall, E. O. (2018). Foodborne pathogens in unpasteurized milk in Sweden. *International Journal of Food Microbiology*, 284, 120–127.

Barreiro, J. R., Ferreira, C. R., Sanvido, G. B., Kostrzewa, M., Maier, T., Wegemann, B., et al. (2010). Short communication: Identification of subclinical cow mastitis pathogens in milk by matrix-assisted laser desorption/ionization time-of-flight mass spectrometry. *Journal of Dairy Science*, 93(12), 5661–5667.

Bassi, D., Cappa, F., Gazzola, S., Orrù, L., & Cocconcellia, P. S. (2017). Biofilm formation on stainless steel by *Streptococcus thermophilus* UC8547 in milk environments ss mediated by the proteinase PrtS. *Applied and Environmental Microbiology*, 83(8), e02840-16.

Bleve, G., Rizzotti, L., Dellaglio, F., & Torriani, S. (2003). Development of reverse transcription (RT)-PCR and real-time RT-PCR assays for rapid detection and quantification of viable yeasts and molds contaminating yogurts and pasteurized food products. *Applied and Environmental Microbiology*, 69(7), 4116–4122.

Boor, K. J., Wiedmann, M., Murphy, S., & Alcaine, S. (2017). A 100-year review: Microbiology and safety of milk handling. *Journal of Dairy Science*, 100, 9933–9951.

Bosward, K. L., House, J. K., Deveridge, A., Mathews, K., & Sheehy, P. A. (2016). Development of a loop-mediated isothermal amplification assay for the detection of *Streptococcus agalactiae* in bovine milk. *Journal of Dairy Science*, 99(3), 2142–2150.

Bottari, B., & Santarelli, M. (2015). Determination of microbial load for different beverages and foodstuff by assessment of intracellular ATP. *Trends in Food Science & Technology*, 44(1), 36–48.

Boyle, T., Njoroge, J. M., Jones, R. L., Jr., & Principato, M. (2010). Detection of staphylococcal enterotoxin B in milk and milk products using immunodiagnostic lateral flow devices. *Journal of AOAC International*, 93(2), 569–576.

Brändle, J., Domig, K. J., & Kneifel, W. (2016). Relevance and analysis of butyric acid producing clostridia in milk and cheese. *Food Control, 67*, 96e113.
Casani, S., Flemming Hansen, K., & Chartier, S. (2015). Interlaboratory collaborative study on a flow cytometry method for lactic acid bacteria quantification in starter cultures, probiotics and fermented milk products according to ISO 19344/IDF 232. *Bulletin of the International Dairy Federation, 478*, 1–48.
Chiao, D., Wey, J., Tsui, P., Lin, F., & Shyu, R. (2013). Comparison of LFA with PCR and RPLA in detecting SEB from isolated clinical strains of *Staphylococcus aureus* and its application in food samples. *Food Chemistry, 141*(3), 1789–1795.
Cho, I. H., & Irudayaraj, J. (2013). Lateral-flow enzyme immunoconcentration for rapid detection of *Listeria monocytogenes*. *Analytical and Bioanalytical Chemistry, 405*(10), 3313–3319.
Commission Regulation. (2004). Regulation (EC) No 853/2004 of the European parliament and of the council of 29 April 2004 laying down specific hygiene rules for food of animal origin. *Official Journal of the European Union* OJ L226; 22.
Delgado, S., Rachid, C. T. C. C., Fernández, E., Rychlik, T., Alegría, Á., Peixoto, R. S., et al. (2013). Diversity of thermophilic bacteria in raw, pasteurized and selectively-cultured milk, as assessed by culturing, PCR-DGGE and pyrosequencing. *Food Microbiology, 36*, 103e111.
Dong, L., Liu, H., Meng, L., Xing, M., Wang, J., Wang, C., et al. (2018). Quantitative PCR coupled with sodium dodecyl sulfate and propidium monoazide for detection of viable *Staphylococcus aureus* in milk. *Journal of Dairy Science, 101*(6), 4936–4943.
Doyle, C. J., Gleeson, D., Jordan, K., Beresford, T. P., Ross, R. P., Fitzgerald, G. F., et al. (2015). Anaerobic sporeformers and their significance with respect to milk and dairy products. *International Journal of Food Microbiology, 197*, 77–87.
Eijlander, R. T., van Hekezen, R., Bienvenue, A., Girard, V., Hoornstra, E., Johnson, S. B., et al. (2019). Spores in dairy—New insights in detection, enumeration and risk assessment. *International Journal of Dairy Technology, 70*, 1–13.
Eisgruber, H., & Reuter, G. (1995). A selective medium for the detection and enumeration of mesophilic sulfite-reducing clostridia in food monitoring programs. *Food Research International, 28*(3), 219–226.
El-sharoud, W. M. (2015). Developing a time and effort-effective, highly sensitive Taq Man probe-based real-time polymerase chain reaction protocol for the detection of *Salmonella* in milk, yoghurt, and cheese. *International Dairy Journal, 40*, 62–66.
European Commission. (2005). *Commission regulation (EC) no 2073/2005 of 15 November 2005 on microbiological criteria for foodstuffs*.
European Commission. (2007). *Commission regulation (EC) no 1441/2007 of 5 December 2007 amending regulation (EC) no 2073/2005 on microbiological criteria for foodstuffs*.
European Commission. (2010). *Commission regulation (EU) no 365/2010 of 28 April 2010 amending regulation (EC) no 2073/2005 on microbiological criteria for foodstuffs as regards Enterobacteriaceae in pasteurized milk and other pasteurized liquid dairy products and* Listeria monocytogenes *in food grade salt*.
European Food Safety Authority & European Centre for Disease Prevention and Control. (2018). The European Union summary report on trends and sources of zoonoses, zoonotic agents and food-borne outbreaks in 2017. *EFSA Journal, 16*(12), 5500.
European Parliament and Council. (2002). *Regulation (EC) no 178/2002 of the European Parliament and of the Council of 28 January 2002 laying down the general principles and requirements of food law, establishing the European Food Safety Authority and laying down procedures in matters of food safety*.
Fernandes, A. C., Duarte, C. M., Cardoso, F. A., Bexiga, R., Cardoso, S., & Freitas, P. P. (2014). Lab-on-chip cytometry based on magnetoresistive sensors for bacteria detection in milk. *Sensors, 14*, 15496–15524.

Flint, S., Bremer, P., Brooks, J., Palmer, J., Sadiq, F. A., Seale, B., et al. (2020). Bacterial fouling in dairy processing. *International Dairy Journal, 101*, 104593.
Food Safety Association of Ireland. (2019). *Guidance note no. 3; guidelines for the interpretation of results of microbiological testing of ready-to-eat foods placed on the market (revision 3)*. Dublin, Ireland: FSAI.
Food Safety Authority of Ireland. (2014). *Guidance note no. 27; Guidance note on the Enforcement of Commission Regulation (EC) no 2073/2005 on microbiological criteria for foodstuffs* (1st ed.). Dublin, Ireland: FSAI.
Forghani, F., Langaee, T., Eskandari, M., Seo, K. H., Chung, M. J., & Oh, D. H. (2015). Rapid detection of viable *Bacillus cereus* emetic and enterotoxic strains in food by coupling propidium monoazide and multiplex PCR (PMA-mPCR). *Food Control, 55*, 151–157.
Franz, C. M. A. P., Stiles, M. E., Schleifer, K. H., & Holzapfel, W. H. (2003). Enterococci in foods—A conundrum for food safety. *International Journal of Food Microbiology, 88*(2–3), 105–122.
Gelsomino, R., Vancanneyt, M., Condon, S., Swings, C. J., & Cogan, T. M. (2002). Enterococcal diversity in the environment of an Irish Cheddar-type cheesemaking factory. *International Journal of Food Microbiology, 71*, 177–188.
Geng, J., Chiron, C., & Combrisson, J. (2014). Rapid and specific enumeration of viable Bifidobacteria in dairy products based on flow cytometry technology: A proof of concept study. *International Dairy Journal, 37*(1), 1–4.
Gleeson, D., O'Connell, A., & Jordan, K. (2013). Review of potential sources and control of thermoduric bacteria in bulk-tank milk. *Irish Journal of Agricultural and Food Research, 52*(2), 217–227.
Gunasekera, T. S., Attfield, P. V., & Veal, D. A. (2000). A flow cytometry method for rapid detection and enumeration of total bacteria in milk. *Applied and Environmental Microbiology, 66*(3), 1228–1232.
Hameed, S., Xiea, L., & Ying, Y. (2018). Conventional and emerging detection techniques for pathogenic bacteria in food science: A review. *Trends in Food Science and Technology, 81*, 61–73.
Han, J., Zhang, L., Hu, L., Xing, K., Lu, X., Huang, Y., et al. (2018). Nanozyme-based lateral flow assay for the sensitive detection of *Escherichia coli* O157:H7 in milk. *Journal of Dairy Science, 101*(7), 5770–5779.
Hennekinne, J. A., Guillier, F., Perelle, S., De Buyser, M. L., Dragacci, S., Krys, S., et al. (2007). Intralaboratory validation according to the EN ISO 16 140 standard of the Vidas SET2 detection kit for use in official controls of staphylococcal enterotoxins in milk products. *Journal of Applied Microbiology, 102*(5), 1261–1272.
Hernández, A., Pérez-Nevado, F., Ruiz-Moyano, S., Serradilla, M. J., Villalobos, M. C., Martín, A., et al. (2018). Spoilage yeasts: What are the sources of contamination of foods and beverages? *International Journal of Food Microbiology, 286*, 98–110.
Hervert, C. J., Alles, A. S., Martin, N. H., Boor, K. J., & Wiedmann, M. (2016). Evaluation of different methods to detect microbial hygiene indicators relevant in the dairy industry. *Journal of Dairy Science, 99*, 7033–7042.
Islam, M. A., Roya, S., Nabia, A., Solaiman, S., Rahman, M., Huq, M., et al. (2018). Microbiological quality assessment of milk at different stages of the dairy value chain in a developing country setting international. *Journal of Food Microbiology, 278*, 11–19.
ISO. (2004). *ISO 21187: Milk—Quantitative determination of bacteriological quality—Guidance for establishing and verifying a conversion relationship between routine method results and anchor method results* (1st ed.). Geneva: ISO.
ISO. (2015). *ISO 19344: Milk and milk products—Starter cultures, probiotics and fermented products—Quantification of lactic acid bacteria by flow cytometry* (1st ed.). Geneva: ISO.

ISO. (2016). *ISO 16140-2: Microbiology of the food chain—method validation-part 2: Protocol for the validation of alternative (proprietary) methods against a reference method* (1st ed.). Geneva: ISO.

Jackson, S. A., Schoeni, J. L., Vegge, C., Pane, M., Stahl, B., Bradley, M., et al. (2019). Improving end-user trust in the quality of commercial probiotic products. *Frontiers in Microbiology, 10*, 1–15. Mar.

Jadhav, S., Gulati, V., Fox, E. M., Karpe, A., Beale, D. J., Sevior, D., et al. (2015). Rapid identification and source-tracking of *Listeria monocytogenes* using MALDI-TOF mass spectrometry. *International Journal of Food Microbiology, 202*, 1–9.

Jadhav, S. R., Shah, R. M., Karpe, A. V., & Morrison, P. D. (2018). Detection of foodborne pathogens using proteomics and metabolomics-based approaches. *Frontiers in Microbiology, 9*, 1–13. December.

Jakobsen, M., & Narvhus, J. (1996). Yeasts and their possible beneficial and negative effects on the quality of dairy products. *International Dairy Journal, 6*, 755–768.

Janganan, T. K., Mullin, N., Tzokov, S. B., Stringer, S., Fagan, R. P., Hobbs, J. K., et al. (2016). Characterization of the spore surface and exosporium proteins of *Clostridium sporogenes*; implications for *Clostridium botulinum* group I strains. *Food Microbiology, 59*, 205–212.

Jin, W., Yamada, K., Ikami, M., Kaji, N., Tokeshi, M., Atsumi, Y., et al. (2013). Application of IgY to sandwich enzyme-linked immunosorbent assays, lateral flow devices, and immunopillar chips for detecting staphylococcal enterotoxins in milk and dairy products. *Journal of Microbiological Methods, 92*(3), 323–331.

Johnson, N., Chang, Z., Almeida, C. B., Michel, M., Iversen, C., & Callanan, M. (2014). Evaluation of indirect impedance for measuring microbial growth in complex food matrices. *Food Microbiology, 42*, 8–13.

Kable, M. E., Srisengfa, Y., Xue, Z., Coates, L. C., & Marco, M. L. (2019). Viable and total bacterial populations undergo equipment and time-dependent shifts during milk processing. *Applied and Environmental Microbiology, 85*(13), e00270-19.

Kennedy, D., & Wilkinson, M. G. (2017). Application of flow cytometry to the detection of pathogenic bacteria. *Current Issues in Molecular Biology, 23*, 21–38.

Khanal, S. M., Anan, S., & Muthukumarappan, K. (2014). Evaluation of high-intensity ultrasonication for the inactivation of endospores of 3 *Bacillus* species in nonfat milk. *Journal of Dairy Science, 97*, 5952–5963.

Kumar, S., Balakrishna, K., & Batra, H. (2008). Enrichment-ELISA for detection of *Salmonella* typhi from food and water samples. *Biomedical and Environmental Sciences, 21*(2), 137–143.

Kumar, S., & Kumar Mondal, K. (2015). Visual detection of *Escherichia coli* contamination in milk and fruit juice using loop-mediated isothermal amplification. *Journal of Food Science and Technology, 52*, 7417–7424. November.

Lavilla, M., Marzo, I., de Luis, R., Perez, M. D., Calvo, M., & Sánchez, L. (2010). Detection of *Clostridium tyrobutyricum* spores using polyclonal antibodies and flow cytometry. *Journal of Applied Microbiology, 108*, 488–498.

Liang, M., Zhang, T., Liu, X., Fan, Y., Xia, S., Xiang, Y., et al. (2015). Development of an indirect competitive enzyme-linked immunosorbent assay based on the multiepitope peptide for the synchronous detection of staphylococcal enterotoxin A and G proteins in milk. *Journal of Food Protection, 78*(2), 362–369.

Lin, S. H., Leong, S. L., Dewan, R. K., Bloomfield, V. A., & Morr, C. V. (1972). Effect of calcium ion on the structure of native bovine casein micelles. *Biochemistry, 11*(10), 1818–1821.

Liu, S., Sui, Z., Wang, B., & Gu, S. (2019). Rapid detection of single viable *Escherichia coli* O157: H7 cells in milk by flow cytometry. *Journal of Food Safety, 39*, 1–9. April.

Loss, G., Apprich, S., Kneifel, W., Mutius, E., & Genuneit, J. (2012). Short communication: Appropriate and alternative methods to determine viable bacterial counts in cow milk samples. *Journal of Dairy Science, 95*(6), 2916–2918.

Luo, D., Huang, X., Mao, Y., Chen, C., Li, F., Xu, H., et al. (2017). Two-step large-volume magnetic separation combined with PCR assay for sensitive detection of *Listeria monocytogenes* in pasteurized milk. *Journal of Dairy Science, 100*(10), 7883–7890.

Madden, R. H., Gordon, A., & Corcionivoschi, N. (2017). Determination of the BactoScan conversion factor for the United Kingdom. *Milk Science International, 70,* 17–19.

Maheux, A. F., Bissonnette, L., Boissinot, M., Bernier, J. L.-V., Bérubé, È, Boudreau, D. K, et al. (2011). Method for rapid and sensitive detection of *Enterococcus* sp. and *Enterococcus faecalis/faecium* cells in potable water samples. *Water Research, 45,* 2342–2354.

Marchin, S., Putaux, J.-L., Pignon, F., & Léonil, J. (2009). Effects of the environmental factors on the casein micelle structure studied by cryo transmission electron microscopy and small-angle x-ray scattering/ultrasmall-angle x-ray scattering. *The Journal of Chemical Physics, 126*(2007), 1–10.

Martin, N. H., Carey, N. R., Murphy, S. C., Wiedmann, M., & Boor, K. J. (2012). A decade of improvement: New York state fluid milk quality. *Journal of Dairy Science, 95,* 7384–7390.

Martin, N. H., Trmčić, A., Hsieh, T.-H., Boor, K., & Wiedmann, M. (2016). Evolving role of coliforms as indicators of unhygienic processing conditions in dairy foods. *Frontiers in Microbiology, 7,* 1549.

McClelland, R. G., & Pinder, A. C. (1994). Detection of *Salmonella typhimunrum* in dairy products with flow cytometry and monoclonal antibodies. *Applied and Environmental Microbiology, 60*(12), 4255–4262.

McHugh, A. J., Feehily, C., Hill, C., & Cotter, P. D. (2017). Detection and enumeration of spore-forming bacteria in powdered dairy products. *Frontiers in Microbiology, 8,* 109.

McHugh, A. J., Feehily, C., Fenelon, M. A., Gleeson, D., Hill, C., & Cotter, P. D. (2020). Tracking the dairy microbiota from farm bulk tank to skimmed milk powder. *Applied and Environmental Science, 5*(2), e00226–20.

Melo, J., Andrew, P. W., & Faleiro, M. L. (2015). *Listeria monocytogenes* in cheese and the dairy environment remains a food safety challenge: The role of stress responses. *Food Research International, 67,* 75–90.

Nyhan, L., Begley, M., Mutel, A., Qu, Y., Johnson, N., & Callanan, M. (2018). Predicting the combinatorial effects of water activity, pH and organic acids on *Listeria* growth in media and complex food matrices. *Food Microbiology, 74,* 75–85.

Parente, E., Ricciardi, A., & Zotta, T. (2020). The microbiota of dairy milk: A review. *International Dairy Journal, 107,* 104714.

Park, H. S., Yang, J., Choi, H. J., & Kim, K. H. (2017). Effective thermal inactivation of the spores of *Bacillus cereus* biofilms using microwave. *Journal of Microbiology and Biotechnology, 27*(7), 1209–1215.

Paul, M., Hekken, D. L., & Brewster, J. D. (2013). Detection and quantitation of *Escherichia coli* O157 in raw milk by direct qPCR. *International Dairy Journal, 32*(2), 53–60.

Poghossian, A., Geissler, H., & Schöning, M. J. (2019). Rapid methods and sensors for milk quality monitoring and spoilage detection. *Biosensors and Bioelectronics, 140,* 111272.

Reineke, K., & Mathys, A. (2020). Endospore inactivation by emerging technologies: A review of target structures and inactivation mechanisms. *Annual Review of Food Science and Technology, 11*(1), 1–1.20.

Rico-Muñoz, E., Samson, R. A., & Houbraken, J. (2019). Mould spoilage of foods and beverages: Using the right methodology. *Food Microbiology, 81,* 51–62.

Roberson, J. R., Fox, L. K., Hancock, D. D., & Besser, T. E. (1992). Evaluation of methods for differentiation of coagulase-positive staphylococci. *Journal of Clinical Microbiology, 30*(12), 3217–3219.

Rohde, A., Hammerl, J. A., Appel, B., Dieckmann, R., & Al Dahouk, S. (2015). FISHing for bacteria in food: A promising tool for the reliable detection of pathogenic bacteria? *Food Microbiology*, *46*, 395–e407.

Sadiq, F. A., Flint, S., & He, Q. (2018). Microbiota of milk powders and the heat resistance and spoilage potential of aerobic spore-forming bacteria. *International Dairy Journal*, *85*, 159–e168.

Seo, K. H., Brackett, R. E., & Frank, J. F. (1998). Rapid detection of *Escherichia coli* O157: H7 using immuno- magnetic flow cytometry in ground beef, apple juice, and milk. *International Journal of Food Microbiology*, *44*, 115–123.

Shan, S., Liu, D., Guo, Q., Wu, S., Chen, R., Luo, K., et al. (2016). Sensitive detection of *Escherichia coli* O157:H7 based on cascade signal amplification in ELISA. *Journal of Dairy Science*, *99*(9), 7025–7032.

Soejima, T., Minami, J., & Iwatsuki, K. (2012). Rapid propidium monoazide PCR assay for the exclusive detection of viable Enterobacteriaceae cells in pasteurized milk. *Journal of Dairy Science*, *95*(7), 3634–3642.

Sohier, D., Pavan, S., Riou, A., Combrisson, J., & Postollec, F. (2014). Evolution of microbiological analytical methods for dairy industry needs. *Frontiers in Microbiology*, *5*, 16.

Somerton, B., Palmer, J., Brooks, J., Smolinski, E., Lindsay, D., & Flint, S. (2012). Influence of cations on growth of thermophilic *Geobacillus* spp. and *Anoxybacillus flavithermus* in planktonic culture. *Applied and Environmental Microbiology*, *78*(7), 2477–2481.

Song, C., Liu, C., Wu, S., Li, H., Guo, H., Yang, B., et al. (2016). Development of a lateral flow colloidal gold immunoassay strip for the simultaneous detection of *Shigella boydii* and *Escherichia coli* O157:H7 in bread, milk and jelly samples. *Food Control*, *59*, 345–351.

Sørhaug, T. (2011). Yeasts and molds. In J. W. Fuquay (Ed.), *Spoilage molds in dairy products. Encyclopedia of dairy sciences* (pp. 780–784). Cambridge, MA, USA: Academic Press.

Sørhaug, T., & Stepaniak, L. (1997). Psychrotrophs and their enzymes in milk and dairy products: Quality aspects. *Trends in Food Science and Technology*, *8*, 35–41.

Suriyarachchi, V. R., & Fleet, G. H. (1981). Occurrence and growth of yeasts in yogurts. *Applied and Environmental Microbiology*, *42*(3), 574–579.

Thomas, A., & Prasad, V. (2014). Thermoduric bacteria in milk- a review. *International Journal of Science and Research*, *3*(6), 2438–2442.

Tirloni, E., Bernardi, C., Drago, S., Stampone, G., Pomilio, F., Cattaneo, P., et al. (2017). Evaluation of a loop-mediated isothermal amplification method for the detection of *Listeria monocytogenes* in dairy food. *Italian Journal of Food Safety*, *6*, 179–184.

Tomáška, M., Suhren, G., Hanuš, O., Walte, H.-G., Slottova, A., & Hofericova, M. (2006). The application of flow cytometry in determining the bacteriological quality of raw sheep's milk in Slovakia. *Le Lait*, *86*, 127–140.

Upadhyay, N., & Nara, S. (2018). Lateral flow assay for rapid detection of *Staphylococcus aureus* enterotoxin A in milk. *Microchemical Journal*, *137*, 435–442.

Van der Mei, H. C., de Vries, J., & Busscher, H. J. (1993). Hydrophobic and electrostatic cell surface properties of thermophilic dairy Streptococci. *Applied and Environmental Microbiology*, *59*(12), 4305–4312.

Vanderhaeghen, W., Pieper, S., Leroy, F., van Coillie, E., Haesebrouck, F., & de Vliegher, S. (2015). Identification, typing, ecology and epidemiology of coagulase negative staphylococci associated with ruminants. *The Veterinary Journal*, *203*, 44–51.

Walstra, P. (1990). On the stability of casein micelles. *Journal of Dairy Science*, *73*, 1965–1979.

Wan, C., Yang, Y., Xu, H., Aguilar, Z. P., Liu, C., & Lai, W. (2012). Development of a propidium monoazide treatment combined with loop-mediated isothermal amplification (PMA-LAMP) assay for rapid detection of viable *Listeria monocytogenes*. *International Journal of Food Science & Technology*, *47*, 2460–2467.

Weenk, G. H., van den Brink, J. A., Struijk, C. B., & Mossel, D. A. A. (1995). Modified methods for the enumeration of spores of mesophilic *Clostridium* species in dried foods. *International Journal of Food Microbiology*, *27*, 185–200.

Wei, Q., Wang, X., Sun, D.-W., & Pu, H. (2019). Rapid detection and control of psychrotrophic microorganisms in cold storage foods: A review. *Trends in Food Science & Technology, 86,* 453–464.

Wilkinson, M. G. (2018). Flow cytometry as a potential method of measuring bacterial viability in probiotic products: A review. *Trends in Food Science & Technology, 78,* 1–10.

Willis, D., Jørgensen, F., Aird, H., Elviss, N., Fox, A., Jenkins, C., et al. (2018). An assessment of the microbiological quality and safety of raw drinking milk on retail sale in England. *Journal of Applied Microbiology, 124*(2), 535–546.

Xue, H., Zhang, B., He, B., Wang, Z., & Chen, C. (2016). Rapid immunochromatographic assay for *Escherichia coli* O157:H7 in bovine milk using IgY labeled by Fe3OM4/Au composite nanoparticles. *Food Science and Technology Research, 22*(1), 53–58.

Yu, S., Yan, L., Wu, X., Li, F., Wang, D., & Xu, H. (2017). Multiplex PCR coupled with propidium monoazide for the detection of viable *Cronobacter sakazakii, Bacillus cereus,* and *Salmonella* spp. in milk and milk products. *Journal of Dairy Science, 100*(10), 7874–7882.

Zhang, Z., Feng, L., Xu, H., Liu, C., Shah, N. P., & Wei, H. (2016). Detection of viable enterotoxin-producing *Bacillus cereus* and analysis of toxigenicity from ready-to-eat foods and infant formula milk powder by multiplex PCR. *Journal of Dairy Science, 99*(2), 1047–1055.

Zhou, P., Xie, G., Liang, T., Yu, B., Aguilar, Z., & Xu, H. (2019). Rapid and quantitative detection of viable emetic *Bacillus cereus* by PMA- qPCR assay in milk. *Molecular and Cellular Probes, 47,* 101437. May.

Ziyaina, M., Rasco, B., & Sablani, S. S. (2020). Rapid methods of microbial detection in dairy products. *Food Control, 110,* 107008.

CHAPTER TWO

The microbiology of red brines

Aharon Oren*
The Institute of Life Sciences, The Hebrew University of Jerusalem, Jerusalem, Israel
*Corresponding author: e-mail address: aharon.oren@mail.huji.ac.il

Contents

1. Introduction	58
2. The organisms	60
2.1 The alga *Dunaliella*	60
2.2 The halophilic archaea	61
2.3 The genus *Salinibacter*	69
2.4 Halophilic fungi	71
2.5 Halophilic protists	71
2.6 Halophilic viruses	72
3. Metabolic interactions between the biota	72
3.1 Glycerol	72
3.2 Dihydroxyacetone	73
3.3 D-Lactate	73
3.4 Pyruvate	74
3.5 Acetate	74
4. The pigments of the red brines	75
4.1 Carotenoids	75
4.2 Retinal pigments	78
4.3 What causes the red color of the brines?	79
5. Natural hypersaline environments: Case studies	80
5.1 Great Salt Lake	80
5.2 The Dead Sea	82
5.3 Alkaline hypersaline lakes	84
6. Commercial salt production facilities	85
7. The role of pigmented microorganisms in salt production	96
8. Concluding remarks	97
Acknowledgments	99
References	99

Abstract

The brines of natural salt lakes with total salt concentrations exceeding 30% are often colored red by dense communities of halophilic microorganisms. Such red brines are found in the north arm of Great Salt Lake, Utah, in the alkaline hypersaline lakes of the African Rift Valley, and in the crystallizer ponds of coastal and inland salterns where salt is produced by evaporation of seawater or some other source of saline water. Red blooms were also reported in the Dead Sea in the past. Different types of pigmented microorganisms may contribute to the coloration of the brines. The most important are the halophilic archaea of the class *Halobacteria* that contain bacterioruberin carotenoids as well as bacteriorhodopsin and other retinal pigments, β-carotene-rich species of the unicellular green algal genus *Dunaliella* and bacteria of the genus *Salinibacter* (class *Rhodothermia*) that contain the carotenoid salinixanthin and the retinal protein xanthorhodopsin. Densities of prokaryotes in red brines often exceed $2-3 \times 10^7$ cells/mL. I here review the information on the biota of the red brines, the interactions between the organisms present, as well as the possible roles of the red halophilic microorganisms in the salt production process and some applied aspects of carotenoids and retinal proteins produced by the different types of halophiles inhabiting the red brines.

1. Introduction

疏卤地为畦陇，而堑围之。引清水注入，久则色赤。待夏秋南风大起，则一夜结，谓之盐南风。如南风不起，则盐失利。

[The ground used to make salt is a pool surrounded by an embankment. They put clear seawater in it. Many days later, the water will be red. If the south wind blows strongly during summer and autumn, the salt can be generated very quickly. If the south wind does not come, there will be no salt.]

> On connoît que le sel se forme quand l'eau rougit; c'est en cet état qu'étant réchauffé par le soleil & par le vent, il se crême de l'épaisseur de verre: alors on le casse, il va au fond, …; il s'y forme en grains gros comme des pois, …

[One knows that the salt is formed when the water becomes red. At this stage, being heated by the sun and the wind, it thickens to the thickness of glass; when one breaks it, it sinks to the bottom, … It is formed as grains as big as peas …]

> One day I rode to a large salt lake, or Salina, which is distant fifteen miles from the town. During the winter it consists of a shallow lake of brine, which in summer is converted into a field of snow-white salt. … Parts of the lake seen from a short distance appeared of a reddish colour, and this, perhaps, was owing to some infusorial animalcula. …

These three quotes show that the occurrence of red brines in salt lakes and salt producing operations is known for many centuries. The text from *Ben Cao Gang Mu*, the famous treatise of Chinese pharmacology by Shizhen (1578; translation: Guowei Qiu) is, as far as I am aware, the oldest written record of such red brines (Oren & Meng, 2019). The second fragment dates from 1765 and is derived from Diderot's Encyclopédie ou Dictionnaire Raisonné des Sciences, des Arts et des Métiers par une société des gens de lettres, 1765 (translation: Aharon Oren). The final one is found in the writings of Charles Darwin, who saw red waters in a coastal hypersaline lake near El Carmen, Patagonia in 1833, one of the stops during the voyage of the Beagle (Darwin, 1839).

In an essay entitled "Red—the magic color for solar salt production," Litchfield (1991) provided a beautiful overview of the appearance of red brines during seawater evaporation, the causes of the red color, and the contribution of the red pigments to the salt production process. Fig. 1 shows an example of such red brines in a commercial salt production facility.

In the following sections, I review the nature of the pigmented halophilic microorganisms inhabiting natural salt lakes and man-made salterns, the interactions between the components of the microbial communities in hypersaline brines, and the possible contribution of the red microbes to the process of salt production by evaporation of seawater.

Fig. 1 Red brines of the crystallizer ponds of the salterns of Salt of the Earth Ltd., Eilat, Red Sea coast of Israel. *Photographs by the author.*

2. The organisms
2.1 The alga *Dunaliella*

Unicellular flagellated green algae of the genus *Dunaliella* (Chlorophyceae) are a characteristic component of the biota of most hypersaline environments. *Dunaliella* species can be found up to the highest salinities (halite saturation), and they are the main or even sole primary producers in saltern crystallizer brines, the Dead Sea, and other environments where the salt concentrations exceed 25–30%. The large red cells of *Dunaliella salina* were first described in saltern ponds in France by Dunal (1838). They appeared in the literature under different names until they were described as a new genus by Teodoresco (1905). The early research on *Dunaliella* was reviewed by Oren (2005), a 100 years after the genus was named. Today approximately 28 species are recognized (González, Gómez, & Polle, 2009), most of which grow at elevated salt concentrations. Fig. 2 shows orange *D. salina* cells in saltern crystallizer brine of Eilat, Israel.

The monograph edited by Ben-Amotz, Polle, and Subba Rao (2009) covers many aspects of the biodiversity, physiology, genomics and biotechnology of the genus *Dunaliella*. The genus is of considerable interest because of its biotechnological applications. Some species (*D. salina*, *D. bardawil*) can

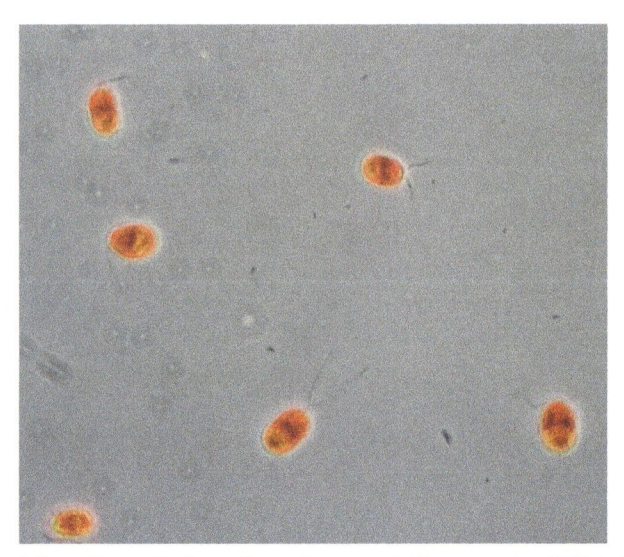

Fig. 2 *Dunaliella salina* cells from the saltern crystallizer ponds of Eilat, Israel. *From Oren, A. (2005). A hundred years of Dunaliella research—1905-2005.* Saline Systems, 1, 2; © Aharon Oren.

under suitable conditions accumulate massive amounts of β-carotene and other carotenoid pigments (Jeffrey & Egeland, 2009). This property is exploited for the commercial production of β-carotene. *Dunaliella* cells contribute to the red coloration of many brines (see Section 4.3). Other species, notably *D. tertiolecta*, found interest in the biodiesel industry. To provide osmotic equilibrium of the cytoplasm with the hypersaline environment, cells accumulate glycerol as an osmotic solute, often to molar intracellular concentrations. Commercial production of glycerol using *Dunaliella* has also been explored.

Red (*D. salina*) and green (*D. viridis*) species abound in the north arm of Great Salt Lake, Utah, and are the main primary producers there (Post, 1977). The formation of red blooms in the Dead Sea in the summers of 1980 and 1992 was preceded by massive development of green *Dunaliella* cells, reaching densities of 8.8×10^3 and 1.5×10^4 cells/mL, respectively (Oren, 1993; Oren, Gurevich, Anati, Barkan, & Luz, 1995; Oren & Shilo, 1982). Due to the toxic concentrations of divalent cations (magnesium, calcium), the waters of the Dead Sea are currently too hostile for development of the alga. Both blooms were triggered by significant dilution of the upper water layers following massive rain floods, combined with the availability of phosphate, which is the limiting nutrient in the Dead Sea (Oren & Shilo, 1982; Oren, Gurevich, Anati, et al., 1995; see also Section 5.2).

Organic compounds produced photosynthetically by *Dunaliella* are the main carbon and energy sources that support development of the heterotrophic communities of halophilic prokaryotes in hypersaline environments, archaea as well as bacteria. Glycerol produced by the alga in massive amounts as an osmotic stabilizer may be one of the principal compounds involved (Elevi Bardavid, Khristo, & Oren, 2008; Oren, 2017; see also Section 3.1).

In saltern crystallizer ponds, β-carotene-rich *D. salina* is typically present in numbers between 10^3 and 10^4 cells/mL, and sometimes even more (Oren, 2009a; Rodriguez-Valera, Ventosa, Juez, & Imhoff, 1985). In spite of the fact that entire hypersaline ecosystems depend on carbon fixed by *Dunaliella* as the main or even the sole primary producer, relatively little is known about the ecology of the different species and their in situ activities (Oren, 2014a).

2.2 The halophilic archaea

Members of the class *Halobacteria*, affiliated with the *Euryarchaeota* phylum, are the main component of the microbial communities in red hypersaline brines.

The *Halobacteria* are a phylogenetically coherent group of salt-requiring prokaryotes (Oren, 2002, 2011, 2014b, 2019b). Until 1986, only two genera were recognized, the rod-shaped *Halobacterium* and the coccoid *Halococcus*. Based on numerical taxonomic analysis of a large number of isolates from saltern crystallizer ponds and other hypersaline environments, the genera *Haloferax* and *Haloarcula* were added (Torreblanca et al., 1986). At the time of writing (June 2020), the class *Halobacteria* encompasses ~70 genera and ~260 species, classified in three orders and six families: the *Halobacteriales* (families *Halobacteriaceae, Haloarculaceae, Halococcaceae*), the *Haloferacales* (families *Haloferacaceae, Halorubraceae*), and the *Natrialbales* (family *Natrialbaceae*) (Gupta, Naushad, & Baker, 2015; Gupta, Naushad, Fabros, & Adeolu, 2016).

Most neutral-pH brines that support massive growth of *Halobacteria* are dominated by Na^+ and Cl^- as the main ions: "thalassohaline" environments, whose ionic composition resembles that of seawater. "Athalassohaline" brines with very different ionic ratios enable growth of specialized organisms such as found, e.g., in the Dead Sea, whose waters currently contain ~2.04 M Mg^{2+} and 0.49 M Ca^{2+} in addition to ~1.29 M Na^+ and 0.21 M K^+. The main anions are Cl^- (6.49 M) and Br^- (0.08 M). These concentrations of magnesium and calcium are too high even for the most divalent cation-tolerant species, but Dead Sea isolates such as *Haloferax volcanii*, *Halorubrum sodomense* and *Halobaculum gomorrense* are markedly tolerant to these ions (Mullakhanbhai & Larsen, 1975; Oren, 1983a; Oren, Gurevich, Gemmell, & Teske, 1995).

Most species of *Halobacteria* prefer near neutral-pH values, but there also are many obligate alkaliphilic types that may cause red coloration of highly alkaline brines such as found in natural soda lakes such as Lake Magadi, Kenya and the Wadi an Natrun lakes in Egypt (Oren, 2013b), and in alkaline saltern evaporation ponds (Gareeb & Setati, 2009; Grant, Grant, Jones, Kato, & Li, 1999; Simachew, Lanzen, Gessesse, & Øvreås, 2016). In such environments, red-pigmented anoxygenic phototrophs of the *Ectothiorhodospira—Halorhodospira* group may also contribute to the color of the brines (Grant & Tindall, 1986; Jannasch, 1957).

Most members of the *Halobacteria* lead an aerobic chemoheterotrophic life style, and they are generally grown in complex media containing amino acids and yeast extract as carbon and energy sources. However, there is considerable metabolic diversity within the group (Andrei, Banciu, & Oren, 2012). Growth of many species is supported by single organic carbon compounds such as simple sugars or organic acids. Some can only be cultivated in nutrient-poor media. Growth media for the flat square *Haloquadratum walsbyi* that is the dominant component of the biota of many saltern crystallizer ponds contain low concentrations of amino acids, pyruvate, and a number of vitamins (Burns et al., 2007). Some members of the *Halobacteria* can also grow anaerobically using nitrate as alternative electron acceptor (Mancinelli & Hochstein, 1986), and/or by reduction of dimethylsulfoxide, trimethylamine-*N*-oxide or fumarate to dimethylsulfide, trimethylamine and succinate, respectively. The strictly anaerobic *Halanaeroarchaeum sulfurireducens*, isolated from sediments and brines collected from hypersaline lakes in the Kulunda Steppe (Altai, Russia), oxidizes acetate or pyruvate with elemental sulfur and a few other sulfur compounds as the electron acceptors (Sorokin, Kublanov, Yakimov, Rijpstra, & Sinninghe Damsté, 2016). Species of *Halobacterium* can grow anaerobically in the dark by fermentation of arginine to ornithine, ammonia and carbon dioxide (Hartmann, Sickinger, & Oesterhelt, 1980). Anaerobic photoheterotrophic growth, using light energy absorbed by the bacteriorhodopsin proton pump, was also shown in some members of the group (Hartmann et al., 1980). *Halorhabdus tiamatea*, a non-pigmented isolate from a deep hypersaline anoxic basin near the bottom of the Red Sea, grows anaerobically; its mode of fermentation was not ascertained. Recently, some unusual ways were discovered used by some members of the *Halobacteria* to making a living, including the aerobic oxidation of carbon monoxide as energy source (King, 2015) and anaerobic lithoheterotrophic growth by *Halodesulfurarchaeum formicicum*, which uses hydrogen or formate as the electron donors and elemental sulfur, thiosulfate or dimethylsulfoxide as the electron acceptors. This species depends on organic nutrients as carbon source (Sorokin et al., 2017).

Nearly all members of the *Halobacteria* are brightly pink-red colored by bacterioruberin carotenoids. Notable exceptions are *Natrialba asiatica*, isolated from beach sand in Japan, and anaerobic representatives such as *Halorhabdus tiamatea* and *Halanaeroarchaeum sulfurireducens*.

In contrast to most other halophilic and halotolerant microorganisms that accumulate organic osmotic solutes within the cells, the red halophilic archaea use potassium and chloride ions to provide intracellular osmotic balance with the outside hypersaline environment (McGenity & Oren, 2012; Oren, 2002, 2013c, 2016a, 2016b). Their intracellular enzymatic machinery is therefore adapted to function in the presence of molar concentrations of salts. Their proteome is characterized by highly acidic, negatively charged proteins that possess a great excess of acidic amino acids (glutamate, aspartate) over basic amino acids (lysine, arginine). Such halophilic proteins generally denature when exposed to low-salt conditions. As a result, high salt concentrations are required for metabolic activity, and often even for structural stability of the cell. The cell wall of most members of the group consists of an acidic glycoprotein that denatures at low salt concentrations. In some members of the *Halobacteria*, trehalose and 2-sulfotrehalose can replace part of the intracellular KCl and contribute to the osmotic balance of the cell (Youssef et al., 2014).

Because of their abundant presence in many red brines and/or their unusual properties, two genera of *Halobacteria* deserve a more in-depth discussion here: *Halorubrum* and *Haloquadratum*.

2.2.1 The genus Halorubrum

Together with the genus *Haloquadratum* that can be easily recognized microscopically but is very difficult to grow, the genus *Halorubrum* is in many cases the main archaeal genus in red saltern brines. *Halorubrum* is the largest genus of the class *Halobacteria*; as of May 2020, 39 species with validly published names had been described. Its species are also easy to grow and readily form colonies on agar plates. Therefore it is not surprising that in cultivation-dependent studies, *Halorubrum* isolates are often the most abundant type recovered. Most species are brightly red by bacterioruberin carotenoids, and they also may contain bacteriorhodopsin and other retinal pigments.

In an early study of the crystallizer ponds at Santa Pola, Alicante, Spain, 16 out of the 17 colonies examined belonged to the genus *Halorubrum*, most of them being closely related to *Halorubrum coriense* (Benlloch et al., 2001). Comparison of the prokaryotic community structure of Spanish Mediterranean (Santa Pola) and Atlantic (Isla Cristina, Huelva) saltern concentrator ponds by a metagenomic approach showed that most sequences were related to the genus *Halorubrum* in Isla Cristina, but to *Haloquadratum* in Santa Pola. Abundance of *Halorubrum* bacteriorhodopsin and halorhodopsin correlates with the abundance of its 16S rRNA gene sequences at Isla Cristina (Fernández et al., 2014). In a molecular analysis of the communities in the solar salterns located in the Odiel marshlands in southwest Spain (33% salinity), *Salinibacter ruber* was identified as the most abundant genus, followed by the archaeal genera *Halorubrum* and *Haloquadratum* (Gómez-Villegas, Vigara, & León, 2018). *Halorubrum*, together with *Haloarcula*, *Haloquadratum* and *Halobacterium*, was a dominant type of archaea of two inland saltern ecosystems in the Alto Vinalopó Valley, Alicante, Spain (Zafrilla, Martínez-Espinosa, Alonso, & Bonete, 2010).

Also in many other salterns around the Mediterranean, *Halorubrum* is a major component of the microbial community. Together with *Haloquadratum*, *Halorubrum* was found most frequently in a study of the

Sfax, Tunisia coastal solar salterns, where flow cytometry and cell sorting were combined with phylogenetic analysis (Trigui et al., 2011). In the crystallizer ponds of the salterns of Sečovlje on the border of Slovenia and Croatia and the salterns at Ston, Croatia, where the communities of halophiles in the crystallizer ponds were not sufficiently dense to impart a red coloration to the brines, most sequences of archaeal 16S rRNA and bacteriorhodopsin genes used as molecular markers grouped within the *Halorubrum* branch (Pašić, Galán Bartual, Poklar Ulrih, Grabnar, & Herzog Velikonja, 2005; Pašić, Ulrih, Črnigoj, Grabnar, & Herzog Velikonja, 2007). *Halorubrum*, in combination with *Haloquadratum*, *Halobacterium*, *Halonotius*, and *Haloarcula*, was also identified as a dominant component of the communities of four salterns in eastern Anatolia (Turkey), based on cultivation-dependent as well as and cultivation-independent methods (Çinar & Mutlu, 2016).

Halorubrum is also a major component of saltern crystallizer communities in Asia and Australia. Strains of *Halorubrum*, together with *Haloarcula*, *Haloferax*, *Halobacterium*, and *Halococcus*, were isolated during the salt harvesting phase in Indian coastal salterns of Goa (Mani, Salgaonkar, & Braganca, 2012) and Tamil Nadu (Manikandan, Kannan, & Pašić, 2009). Cultivation-independent, 16S rRNA gene-based characterization of the microbial diversity in a Korean saltern showed that the majority (56%) was affiliated with the *Halorubrum* group (Park, Kang, & Rhee, 2006). The genera most frequently cultivated from Australian crystallizer ponds were *Haloferax*, *Halorubrum*, and *Natronomonas* (Burns, Camakaris, Janssen, & Dyall-Smith, 2004a). Very similar sequences of *Halorubrum* and *Haloquadratum* were found in three, geographically distant, Australian salterns (Oh, Porter, Russ, Burns, & Dyall-Smith, 2009). *Halorubrum* sequences were also abundantly found together with *Haloquadratum* and other phylotypes in the red brines of the natural hypersaline Lake Tyrrell, Victoria, Australia (Podell, Ugalde, Narasingarao, Banfield, & Heidelberg, 2013). Cultivation-dependent, as well as cultivation-independent studies of the salterns at San Diego, California, USA and Guerrero Negro, Baja California, Mexico also showed abundance of *Halorubrum* (Bidle et al., 2005; Sabet, Diallo, Hays, Jung, & Dillon, 2009).

A study of the *Halorubrum* in Spanish salterns, involving phylogenetic and genome fingerprinting analysis, shows that its populations show rapidly changing variations. Sequence data from multiple loci identified many closely and more distantly related strains belonging to the genera *Halorubrum* and *Haloarcula*. This variation in genome structure proves that accumulation of genomic variation is rapid, even between identical multilocus sequence analysis haplotypes (Mohan et al., 2014). Genes for the glycerol metabolism gene dihydroxyacetone kinase in the microbial communities inhabiting solar salterns at different geographical locations may to a large extent belong to members of the genus *Halorubrum* (Moller & Liang, 2017).

2.2.2 The genus Haloquadratum

Today it is clear that the square archaeon *Haloquadratum walsbyi* (family *Haloferacaceae*, order *Haloferacales*) is the dominant archaeon in many red brines, both in natural salt lakes and in solar saltern crystallizer ponds (Ventosa, de la Haba, Sánchez-Porro, & Papke, 2015). It is somewhat surprising that the existence of this numerically so abundant and morphologically so unusual organism was documented so late. It was first reported from a small pigmented coastal brine pool on the Sinai Peninsula, Egypt, in 1980 by Walsby (1980). Fluorescence in situ hybridization analysis of the microbial community inhabiting the crystallizer ponds of the salterns of Santa Pola, Alicante, Spain showed that the dominant square or rectangular type of prokaryote possessed the phylotype, then designated SPhT, known already for some time as the most abundant phylotype in the ponds (Antón, Llobet-Brossa, Rodríguez-Valera, & Amann, 1999). The dominance of these flat square gas-vacuolated cells in the red brines of the Eilat, Israel salterns enabled the assessment of the polar lipid composition of these cells: its glycolipid was found to be chromatographically identical with the sulfated diglycosyl diether lipid found as the major glycolipid in the genus *Haloferax* and in other related genera (Oren, Duker, & Ritter, 1996).

In spite of many efforts to cultivate this interesting organism, success was reported only 24 years after its discovery. In 2004, two groups independently succeeded in isolating this elusive prokaryote. One culture was obtained from coastal salterns in Victoria, Australia (Burns, Camakaris, Janssen, & Dyall-Smith, 2004b), and another from the Alicante salterns in Spain (Bolhuis, te Poele, & Rodriguez-Valera, 2004). Its genome sequence was published soon after (Bolhuis et al., 2006). A formal description of the organism followed in 2007, and it was named *Haloquadratum walsbyi* (the salty square of Walsby) (Burns et al., 2007). It grows optimally at 23–30% salt. The organism can tolerate very high magnesium concentrations ($>2\,M$ $MgCl_2$ in the presence of $3.3\,M$ NaCl). The cells contain gas vesicles and granules of the storage material poly-β-hydroxyalkanoate.

Presence of *Haloquadratum* in brines can be easily assessed, both thanks to their unique morphology and their unique phylotype, distantly related to other members of the *Haloferacaceae* family. Its abundant presence was reported in hypersaline brines around the world. These include coastal salterns at the Mediterranean coast of Spain (Benlloch et al., 2002; Ghai et al., 2011; Øvreås, Daae, Torsvik, & Rodríguez-Valera, 2003), Isla Cristina, Huelva, close to the Atlantic coast of Spain (Fernández et al., 2014), salterns of the Odiel marshlands (SW Spain) where *Salinibacter* is the most abundant genus, followed by the archaeal genera *Halorubrum* and *Haloquadratum* (Gómez-Villegas et al., 2018), salterns on Mallorca (Balearic Islands, Spain) (Viver et al., 2019), and two inland saltern ecosystems in the Alto Vinalopó Valley, Spain (Zafrilla et al., 2010). Abundance of *Haloquadratum* was reported from other sites around the Mediterranean: the salterns of Sfax (Tunisia) (Boujelben, Gomariz, et al., 2012; Trigui et al., 2011), salterns in eastern Anatolia (Turkey) (Çinar & Mutlu, 2016), and Tuz Lake, a hypersaline inland lake in Turkey (Mutlu et al., 2008). In the traditionally operated Adriatic salterns of Sečovlje, however, *Haloquadratum* was not a dominant genus (Pašić et al., 2005).

In Australian coastal salterns, *Haloquadratum*-related 16S rRNA gene sequences very similar to that of the *H. walsbyi* type strain were found at three geographically distant sites (Oh et al., 2009). *Haloquadratum* is also abundant in the inland hypersaline Lake Tyrrell, Victoria (Podell et al., 2014). It was also found in environmental genomics and metagenomics studies of the microbial communities in salterns in Puerto Rico (Couto-Rodríguez & Montalvo-Rodríguez, 2019), Argentina (Di Meglio et al., 2016), and its sequences dominated the archaeal assemblage of the salterns of Maras, located 3380 m above sea level in the Peruvian Andes (Maturrano, Santos, Rosselló-Mora, & Antón, 2006).

Comparison of the genome sequences of the Australian and the Spanish strain showed a large degree of similarity. The chromosomes are 3.1 Mb in size. Many inferred deletions were associated with short direct repeats (4–20 bp). Deletion-coupled insertions show that *Haloquadratum walsbyi* evolves by uptake and integration of foreign DNA, probably originating from close relatives. Change is also driven by mobile genetic elements (Dyall-Smith et al., 2011). A study of the intraspecies diversity of *Haloquadratum* in a Spanish saltern crystallizer pond indicated presence of a large pool of accessory genes (Legault et al., 2006).

Haloquadratum is a photoheterotroph that can obtain energy from light absorbed by bacteriorhodopsin, which can be present in large amounts in natural communities in saltern brines (Corcelli, 2014). Its recommended growth medium contains low concentrations of peptone (Oxoid), yeast extract and pyruvate as organic carbon sources (Burns et al., 2007). Fluorescence in situ hybridization experiments combined with autoradiography, using samples from Spanish salterns, showed that amino acids are taken up and incorporated into cell material. Acetate, a substrate that did not stimulate growth (Burns et al., 2007), was also incorporated (Rosselló-Mora, Lee, Antón, & Wagner, 2003).

Walsby (1980) first recognized the flat square *Haloquadratum* cells in the sample he collected from a coastal brine pool thanks to the presence of many gas vesicles in the cells. *Haloquadratum* is not the only gas-vacuolated member of the *Halobacteria*, other examples are *Halobacterium salinarum*, *Haloferax mediterranei*, and *Haloplanus natans*. It is often assumed that these halophilic archaea may use the gas vesicles to float to the surface of the brine, where they have better access to oxygen and (for those species that possess retinal pigments) to light. However, experimental evidence as well as theoretical considerations make it clear that for *Haloquadratum* the cells' content of gas vesicles is insufficient to provide positive buoyancy to make the cells rapidly float to the surface. An alternative hypothesis is that the gas vesicles, located mainly close to the cell periphery, may aid the cells to position themselves parallel to the surface, thereby increasing the efficiency of light harvesting (Oren, Pri-El, Shapiro, & Siboni, 2006).

2.2.3 The Nanohaloarchaea

The Nanohaloarchaeota are a recently discovered group of very small extremely halophilic archaea found often in large numbers ($>10^6$ cells/mL) in natural salt lakes and in saltern brines. They were first detected during cultivation-independent studies of the biota of an East-African alkaline saltern (Grant et al., 1999). No members of the group are yet available in culture, but genomes of a number of *Candidatus* genera have been retrieved from metagenomics studies of the biota of red brines of salterns and salt lakes. These include *Candidatus* Nanosalinicola and *Candidatus* Nanosalina from Lake Tyrrell, Australia (Narasingarao et al., 2011; Podell et al., 2014) and *Candidatus* Haloredivivus from the salterns near Alicante, Spain (Ghai et al., 2011; Gomariz et al., 2015). Sequences related to these *Candidatus* genera were abundantly found in Argentinian salterns (Di Meglio et al., 2016) and in many other brines and hypersaline sediments worldwide (Mora-Ruiz et al., 2018). Rhodopsin genes were identified in all Nanohaloarchaeota genomes, and it is therefore probable that they lead a photoheterotrophic life style. Because of the presence of retinal proteins they may contribute to the pigmentation of brines. However, as long as these Nanohaloarchaeota cannot be studied in culture, no further information about their pigmentation is available.

2.3 The genus *Salinibacter*

The abundant presence of an extremely halophilic representative of the bacterial domain in saltern crystallizer ponds was first indicated during fluorescence in situ hybridization analysis of the prokaryotic communities in the salterns near Alicante, Spain. In addition to the many cells with flat square morphology associated with the archaeon *Haloquadratum walsbyi*, slightly curved rods were found that bound bacteria-specific 16S rRNA targeted fluorescent probes (Antón et al., 1999). This type of organism makes up between 5% and 25% of the prokaryotes in many crystallizer ponds (Antón, Rosselló-Mora, Rodríguez Valera, & Amann, 2000). Soon after its discovery the organism was brought into culture. It was described as a new genus and species, *Salinibacter ruber* (Antón et al., 2002). *Salinibacter* was originally placed in the *Bacteroidetes* phylum, but it is current classified in the class *Rhodothermia* in a newly proposed phylum, the *Rhodothermaeota* (Munoz, Rosselló-Móra, & Amann, 2016). It grows optimally at 20–30% salt, and requires at least 15% salt for growth. The pigmentation of *Salinibacter* is mainly due to the presence of salinixanthin, a C_{40}-carotenoid acyl glycoside (Lutnæs, Oren, & Liaaen-Jensen, 2002). In addition, retinal pigments are present. Salinixanthin acts as a light antenna that transfers energy to the retinal group of xanthorhodopsin, a bacteriorhodopsin-like light-driven proton pump (Balashov et al., 2005; Balashov & Lanyi, 2007).

A second species of the genus, *Salinibacter altiplanensis*, was recently described (Viver et al., 2018). Two more species earlier classified in the genus, *S. iranicus* and *S. luteus*, were later transferred to a new genus, *Salinivenus*.

Salinibacter shares many properties with the extremely halophilic archaea of the class *Halobacteria*, including the use of KCl rather than organic solutes to provide osmotic balance, a large excess of acidic amino acids in most proteins, and the possession of retinal pigments. Some of this resemblance may be due to horizontal gene transfer (Mongodin et al., 2005; Oren, 2008, 2013a, 2019a; Oren et al., 2004). *Salinibacter* and its relatives contain an unusual type of sulfonolipids (halocapnine derivatives) (Corcelli et al., 2004; Oren, 2013a).

Presence of *Salinibacter* in the microbial communities of hypersaline brines can be ascertained by 16S rRNA sequence-based techniques (clone libraries, fluorescence in situ hybridization), metagenomics (Ventosa et al., 2015), and by analysis of lipids, based on the presence of the unique sulfonolipid that can be used as a biomarker (Lopalco, Lobasso, Baronio, Angelini, & Corcelli, 2011). Salinixanthin, the carotenoid pigment of *Salinibacter*, also is a convenient biomarker: Analysis of the pigments extracted from Spanish solar saltern crystallizer ponds showed that 5–7.5% of the total prokaryotic pigment absorbance could be attributed to this pigment (Oren & Rodríguez-Valera, 2001).

Cultivation-dependent methods can also be applied: plating samples on suitable high-salt media in the presence of bacitracin to inhibit growth of most members of the *Halobacteria* proved to be an easy method for the isolation of *Salinibacter* from diverse environments. Also anisomycin can be used in selective media for the enrichment of *Salinibacter* with exclusion of the halophilic archaea (Elevi Bardavid et al., 2007).

There are reports of the high abundance of *Salinibacter* in saltern crystallizer brines in Spain (Antón et al., 2008; Benlloch et al., 2002; Gomariz et al., 2015; Gómez-Villegas et al., 2018; Øvreås et al., 2003; Viver et al., 2015), Argentina (Di Meglio et al., 2016), India (Manikandan et al., 2009), Turkey (Mutlu & Güven, 2015), Baja California, Mexico (Sabet et al., 2009), Italy (Lopalco et al., 2011), Tunisia (Trigui et al., 2011), and Israel (Elevi Bardavid et al., 2007). *Salinibacter* was also detected in the red brines of Lake Tyrrell, Victoria, Australia (Podell et al., 2014), Tuz Lake in Turkey (Mutlu et al., 2008), and in salty ponds around the Araruama Lagoon, a coastal lagoon in Brazil (Clementino et al., 2008).

Salinibacter ruber isolates from different sites worldwide are highly similar, but there is some degree of microdiversity. Thus, metagenomic islands—highly variable regions with atypically low GC content, low coding density, high numbers of pseudogenes and short hypothetical proteins, were detected among co-occurring *Salinibacter* cells (Pašić et al., 2009). There also is evidence for biogeographic isolation of strains distinguished by means of characteristic metabolites, but differences are quantitative rather than qualitative (Rosselló-Mora et al., 2008).

Early experiments to elucidate the nutrition of *Salinibacter* in situ in Spanish saltern brines used microautoradiography combined fluorescence in situ hybridization following addition of different radiolabeled substrates. No in situ uptake of amino acids or glycerol could be demonstrated (Rosselló-Mora et al., 2003). However, laboratory studies of *Salinibacter* cultures proved that glycerol, a product of *Dunaliella* photosynthesis, is readily taken up by *Salinibacter* and stimulates its growth (Elevi Bardavid et al., 2008; Sher, Elevi, Mana, & Oren, 2004). Metagenomic studies showed that genes encoding glycerol-3-phosphate dehydrogenase are abundant in *Salinibacter*, suggesting an active glycerol metabolism (Moller & Liang, 2017). Part of the glycerol is incompletely oxidized to dihydroxyacetone, a compound that is a preferred substrate of *Haloquadratum* (Elevi Bardavid et al., 2008; Elevi Bardavid & Oren, 2008).

2.4 Halophilic fungi

Fungi were never shown to be a dominant component in the biota of hypersaline brines. Still, they were found in all hypersaline ecosystems where a targeted search was made. Melanized, black fungi are among the most abundant types encountered, and these may even have some impact on the optical properties of the brines.

Occurrence of melanized halophilic fungi in hypersaline environments was first described from the crystallizers ponds of the Adriatic salterns at Sečovlje, Slovenia. Species of black, yeast-like hyphomycetes encountered are *Hortaea werneckii, Phaeotheca triangularis, Trimmatostroma salinum, Aureobasidium pullulans*, together with some *Cladosporium* species. Hypersaline brines may be the natural ecological niche of *H. werneckii, P. triangularis* and *T. salinum* (Gunde-Cimerman, Zalar, de Hoog, & Plemenitaş, 2000). Melanized fungi have been isolated from hypersaline waters on three continents, indicating that they are present globally in hypersaline waters of man-made salterns (Butinar, Sonjak, Zalar, Plemenitaş, & Gunde-Cimerman, 2005; Diaz-Munoz & Montalvo-Rodriguez, 2005).

2.5 Halophilic protists

Heterotrophic protists, including ciliates and flagellates, are important in the regulation of prokaryotic cell densities at intermediate salinities in saltern ponds and other hypersaline environments (Elloumi, Guermazi, Ayadi, Bouain, & Aleya, 2009; Filker, Gimmler, Dunthorn, Mahé, & Stoeck, 2015; Park & Simpson, 2015). However, there only are few reports of active flagellates grazing on prokaryotes in high salinity waters (>30% salt) of solar salterns. In a Korean saltern crystallizer pond, grazing by heterotrophic nanoflagellates ($7-28 \times 10^3$ cells/mL) was documented (Park, Kim, Choi, & Cho, 2003). One such bacteriovorous nanoflagellate, *Halocafeteria seosinensis* (Bicosoecida), was isolated and characterized. It grew optimally at 15% salt, but tolerated >35% (Park, Cho, & Simpson, 2006). It is widespread in hypersaline environments (Park & Simpson, 2015). Another halophilic heterolobosean flagellate, *Aurem hypersalina*, isolated from brine of 34.2% salinity, grew well at 10–20% salt (Jhin & Park, 2018).

The possibility to use the halophilic ciliate *Fabrea salina* to control excessive development of *Dunaliella salina* in solar salterns was explored (Hong & Choi, 2015). This study was performed at ~9% salinity, a value much lower than the salinities at which red brines are found, and there is no evidence that grazing by *Fabrea* can control *Dunaliella* communities at the highest salinities.

2.6 Halophilic viruses

At the highest salinities (>25% salt), viral lysis appears to be quantitatively much more important than grazing by heterotrophic protists (Guixa-Boixareu, Calderón-Paz, Heldal, Bratbak, & Pedrós-Alió, 1996; Guixa-Boixereu, Lysnes, & Pedrós-Alió, 1999; Pedrós-Alió et al., 2000) An electron micrograph of *Haloquadratum* collected from a Spanish saltern shows ~200 virus particles inside a single cell (Guixa-Boixareu et al., 1996). Cas/CRISPR systems were detected in the genome of *Haloquadratum walsbyi* C23, the type strain that originated from Australia; except for a few spacer remnants no such systems were detected in the Spanish isolate HBSQ001 (Dyall-Smith et al., 2011). In crystallizer ponds in Tunisia (34–36% salt), virus-like particles were 49–54-times as numerous as prokaryotic cells (Boujelben, Yarza, et al., 2012).

In the Dead Sea, viruses attacking halophilic archaea may have been responsible for the decline of the community after a bloom: between 0.9 and 7.3×10^7 virus-like particles/mL were enumerated in the upper 20 m of the water column in October 1994 during the decline of the archaeal bloom that had developed in 1992 (Oren, Bratbak, & Heldal, 1997).

A review on the different approaches to the study of viral communities in hypersaline environments presents further information (Santos et al., 2012).

3. Metabolic interactions between the biota

To understand the metabolic interactions among the components of the microbial consortia that inhabit red brines, there are a few key compounds of special importance. These include glycerol, dihydroxyacetone, D-lactate, pyruvate, and acetate.

3.1 Glycerol

The main or only primary producer in saltern crystallizer ponds and many natural salt lakes at or approaching salt saturation is the unicellular green alga *Dunaliella salina* and other species of the *Dunaliella* genus. *Dunaliella* excludes salts from its cytoplasm, and instead it accumulates photosynthetically produced glycerol as its osmotic solute. Use of *Dunaliella* has even been explored for the commercial production of glycerol. As *Dunaliella* cells contain molar concentrations of glycerol intracellularly, this compound is expected to become available as one of the main organic compounds to the heterotrophic communities of archaea and bacteria in hypersaline ecosystems.

A literature survey of the properties of the halophilic archaea of the class *Halobacteria* shows that the growth of many, but not of all species, is stimulated by glycerol (Oren, 2017). Cultures of *Salinibacter ruber* also take up glycerol, and addition of glycerol leads to increased growth yields. Use of glycerol by *Salinibacter* depends on the activity of an inducible glycerol kinase (Sher et al., 2004). In spite of the widespread use of glycerol as a growth substrate by many halophilic prokaryotes, archaeal as well as bacterial, it is surprising that none of the components of the microbial community in the Santa Pola, Alicante, Spain saltern crystallizer ponds, including *Salinibacter*, showed incorporation of radiolabeled glycerol in experiments in which fluorescence in situ hybridization was combined with autoradiogaphy (Rosselló-Mora et al., 2003).

3.2 Dihydroxyacetone

When cultures of *Salinibacter ruber* are amended with glycerol, the major part of the glycerol is converted to cell material or respired to carbon dioxide. However, up to 20% of the glycerol added is converted to a soluble compound that accumulates in the medium (Sher et al., 2004). This compound was identified as dihydroxyacetone, a product of the partial oxidation of glycerol. An efficient uptake system for dihydroxyacetone uptake was identified in the genome of *Haloquadratum*, one of the main components of the microbial consortium in most saltern crystallizer ponds and other hypersaline ecosystems (Bolhuis et al., 2006). Therefore, degradation of glycerol that originates from *Dunaliella* may involve dihydroxyacetone as an intermediate, so that part of the glycerol is first converted to dihydroxyacetone, which can then be further assimilated or respired by *Haloquadratum* and other members of the *Halobacteria* (Elevi Bardavid & Oren, 2008). Use of dihydroxyacetone by the heterotrophic community in saltern crytallizer ponds of the salterns of Eilat, Israel, was demonstrated by the use of an optical oxygen electrode to study the heterotrophic activity in the brine in the dark: addition of glycerol as well as dihydroxyacetone enhanced respiration rates (Oren, 2016c).

3.3 D-Lactate

Many members of the class *Halobacteria* produce acids when grown in the presence of carbohydrates, sugar alcohols and similar compounds such as glycerol. Members of the genera *Haloferax* and *Haloarcula* are among the best-known organisms that acidify the growth medium in the presence of sugars and similar compounds. Studies in the early 1970s identified the compounds formed from glucose by *Halorubrum saccharovorum* as pyruvate and acetate (Tomlinson & Hochstein, 1972). D-lactate and acetate were identified in cultures of *Haloferax* and *Haloarcula* spp. incubated in the presence of glycerol (Oren & Gurevich, 1994). Radiolabeling experiments showed that lactate is also formed when red saltern crystallizer brines of the Eilat, Israel salterns were supplemented with ^{14}C-labeled glycerol at concentrations of 1.5–3 μM. After depletion of the added glycerol, the lactate was further degraded (Oren & Gurevich, 1994).

3.4 Pyruvate

Many halophilic archaea of the class *Halobacteria* use pyruvate as one of their preferred growth substrates (Oren, 2015). A notable example is the square archaeon *Haloquadratum walsbyi*: pyruvate is an important component of the media that support growth of the few isolates now in culture (Bolhuis et al., 2004; Burns et al., 2004b). Pyruvate is also one of the very few compounds that enable growth of *Halosimplex carlsbadense* (Vreeland et al., 2002). Like glycerol and dihydroxyacetone, pyruvate stimulated the community respiration of crystallizer brines of the salterns in Eilat, Israel (Oren, 2016c). On the other hand, pyruvate can also be produced by some halophilic archaea: it is excreted by *Halorubrum saccharovorum* and by *Haloarcula* spp. in the presence of glucose (Oren & Gurevich, 1994; Tomlinson & Hochstein, 1972). Upon incubation of Dead Sea water (November 1993, with 1.2×10^7 prokaryotes/mL, a remnant of the red bloom that developed in the lake in 1992) with 1.5–3 μM radiolabeled glycerol, radioactive pyruvate was detected together with labeled lactate and acetate; after depletion of the added glycerol, the pyruvate rapidly disappeared. No pyruvate accumulation was demonstrated in similar experiments performed with saltern crystallizer brine from Eilat (Oren & Gurevich, 1994).

3.5 Acetate

Acetate is another compound excreted by many members of the *Halobacteria* class when incubated with sugars or glycerol (Oren & Gurevich, 1994; Tomlinson & Hochstein, 1972). Some (e.g., many *Haloferax* species) can also grow on acetate, either as single substrate or in combination with other carbon sources. Microautoradiography experiments in combination with fluorescent in situ hybridization in which Alicante saltern brine samples were incubated with labeled acetate showed that the flat square *Haloquadratum* cells took up acetate (Rosselló-Mora et al., 2003), a substrate that in cultivation experiments did not stimulate its growth (Burns et al., 2007).

In accordance with the results of cultivation experiments (Antón et al., 2002), *Salinibacter ruber* in the saltern crystallizer brine community did not show incorporation of labeled acetate in the microautoradiography experiments (Rosselló-Mora et al., 2003). Following supplementation of Eilat saltern brine or Dead Sea water with radiolabeled glycerol, labeled acetate was formed by the communities of halophilic microorganisms; this acetate was only slowly degraded after the depletion of glycerol (Oren & Gurevich, 1994). Acetate turnover rates were estimated to be very slow in such brines, being in the order of weeks to months (Oren, 1995).

4. The pigments of the red brines

The main pigments responsible for the color of the red brines are carotenoids. Carotenoids are formed by many microorganisms, and they have important functions in protecting the cells against life under stressful conditions. Being effective scavengers of free radicals, they facilitate life in the presence of oxygen and high light intensities. *Dunaliella salina* and some other *Dunaliella* species produce massive amounts of the C_{40} carotenoid β-carotene, members of the *Halobacteria* are pink-red because of the C_{50} carotenoid α-bacterioruberin and derivatives, and the C_{40}-carotenoid acyl glycoside salinixanthin is mainly responsible for the orange color of *Salinibacter*. The structures of these compounds are shown in Fig. 3. Membrane-bound retinal-containing proteins such as bacteriorhodopsin and halorhodopsin can also contribute to the pigmentation of the cells.

4.1 Carotenoids
4.1.1 Dunaliella carotenoids
The genus *Dunaliella* contains nearly 30 species, living in marine as well as hypersaline environments (Ben-Amotz et al., 2009), and its taxonomy is currently problematic (Polle, Jin, & Ben-Amotz, 2020). As in other members of the Chlorophyceae, the photosynthetic system is based on chlorophyll *a* and chlorophyll *b*, accompanied by different carotenoids, including lutein and β-carotene. The carotenoids are related as intermediates of the same pathway, and accumulation of different intermediates depends upon the conditions (Jeffrey & Egeland, 2009). An HPLC chromatogram of an extract of biomass collected from the Dead Sea in May 1992, during a bloom of green *Dunaliella* cells and the beginning of the development of a bloom of *Halobacteria* that contain C_{50} bacterioruberin carotenoids, shows all these components (Oren, Gurevich, Anati, et al., 1995).

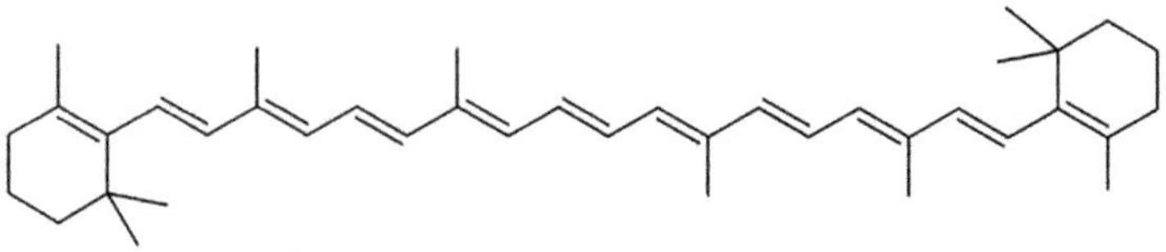

β-Carotene

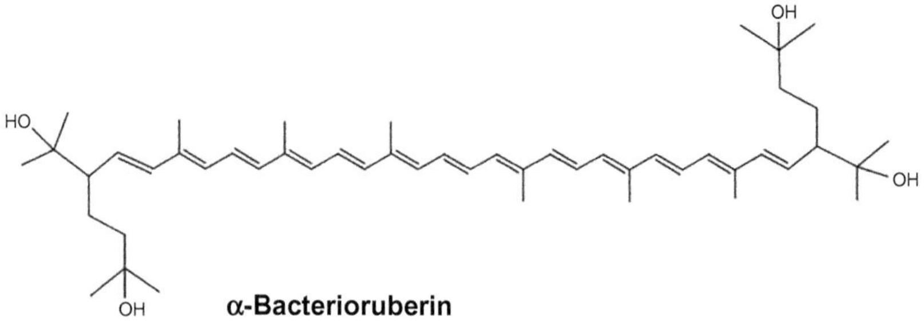

α-Bacterioruberin

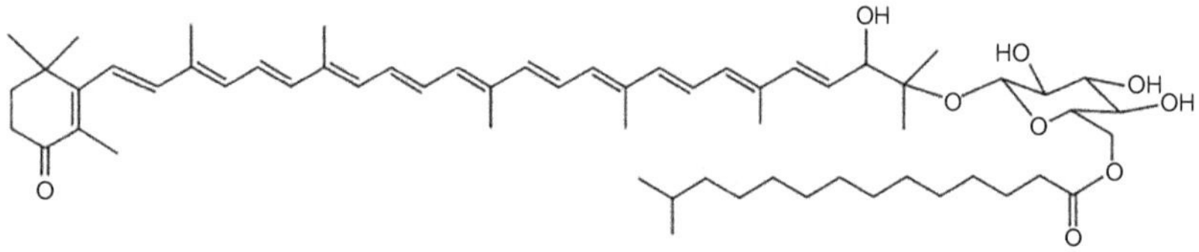

Salinixanthin

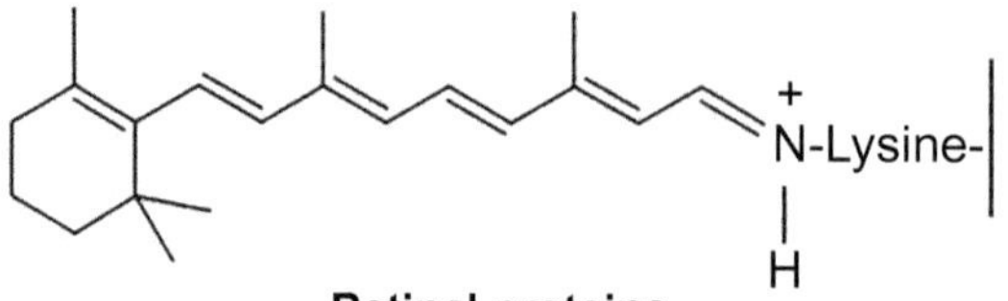

Retinal proteins

Fig. 3 The chemical structures of all-*trans* β-carotene (one of the carotenoids accumulated by *Dunaliella* species), α-bacterioruberin (the main carotenoid of the red halophilic archaea of the class *Halobacteria*), salinixanthin of *Salinibacter ruber*, and the retinal group bound by a Schiff base to a lysine residue in bacteriorhodopsin, halorhodopsin and other retinal proteins.

A few *Dunaliella* species can under suitable conditions produce massive amounts of β-carotene, even exceeding 10% of the cells' dry weight. As a result, the cells become brightly orange colored. *D. salina* and *D. bardawil* are well known for their ability to accumulate β-carotene, which is exploited industrially (Ben-Amotz et al., 2009).

Massive production of β-carotene in orange strains of *Dunaliella* is induced by stressful conditions: high light intensity, high salt concentration, nitrate deficiency and extreme temperatures. The β-carotene is accumulated in globules of ∼150 nm diameter located in the interthylakoid spaces of the chloroplast. Up to half of the β-carotene can be present as the 9-*cis* isomer rather than as all-*trans* β-carotene. The higher the light intensity to which the cells are exposed during growth, the higher the ratio of the 9-*cis* to the all-*trans* isomer. The β-carotene globules may serve to protect the cells against injury by the high intensity irradiation to which this alga is often exposed in its natural habitat (Ben-Amotz, Katz, & Avron, 1982; Ben-Amotz, Lers, & Avron, 1988). A recent study in which quick-freeze deep-etch electron microscopy was used to visualize cellular structures in *D. salina* showed that orange cells have greatly reduced amounts of thylakoid membranes. The β-carotene globules are in contact with the thylakoid membranes, indicating that exchange of molecules between the β-carotene globules and the thylakoid membranes may be possible. β-Carotene globule duplets were observed, possibly as an intermediate stage in the formation of the globules (Polle, Roth, Ben-Amotz, & Goodenough, 2020).

β-Carotene has its absorption maximum at 450 nm, with a minor peak at ∼480 nm. When biomass of saltern crystallizer ponds is collected by filtration and the filters are extracted with organic solvents, the extract is generally yellow-orange colored, and the absorption spectrum is dominated by the β-carotene peak (Oren & Dubinsky, 1994; supplementary fig. 6 in Viver et al., 2019).

4.1.2 Haloarchaeal carotenoids

The pink-red color of the members of the archaeal class *Halobacteria* is mainly due to the presence of the open-chain C_{50} carotenoid α-bacterioruberin and minor amounts of different anhydro derivatives (Kelly, Norgård, & Liaaen-Jensen, 1970; Kushwaha, Gochnauer, Kushner, & Kates, 1974; Kushwaha, Kramer, & Kates, 1975). Their absorption spectrum is dominated by a peak near 496 nm, a minor peak at ∼530 nm, and a shoulder at ∼470 nm. These are the pigments found in red pellets obtained by centrifugation of Dead Sea water during studies performed in 1963–1963 (a period in which no visible red blooms were reported) (Kaplan & Friedmann, 1970) and in 1992 during the massive development of red archaea in the lake (Oren, Gurevich, Anati, et al., 1995).

4.1.3 Salinibacter carotenoids

Cultures of *Salinibacter ruber* are more orange and less pink colored than those of the halophilic archaea. Absorption spectra of pigment extracts in organic solvents show a peak at ~482 nm and a broad shoulder around 506–510 nm. A single carotenoid, named salinixanthin, was found to make up >96% of the cell's carotenoids. It is a C_{40}-carotenoid acyl glycoside first identified in this organism (Lutnæs et al., 2002). Its structure is shown in Fig. 2.

Salinixanthin can be easily separated by HPLC from the archaeal bacterioruberin pigments and from β-carotene and other *Dunaliella*-derived carotenoids. This enables the assessment of the relative abundance of *Salinibacter*-derived salinixanthin and the carotenoids produced by other components of the microbial communities in hypersaline brines. Thus, 5–7.5% of the total prokaryotic pigment absorbance in extracts of cell pellets collected by centrifugation of brine from Spanish saltern crystallizer ponds could be attributed to salinixanthin, while only traces were found, if at all, in similar extracts of biomass collected from the salterns of Eilat, Israel, and Newark, CA, USA (Oren & Rodríguez-Valera, 2001).

4.2 Retinal pigments

Many halophilic prokaryotes possess membrane-bound retinal proteins. After the light-driven proton pump bacteriorhodopsin was discovered in a *Halobacterium* in the early 1970s (Oesterhelt & Stoeckenius, 1971, 1973), retinal proteins with other functions were also found: the light-driven inward chloride pump halorhodopsin and sensory rhodopsins used as light sensors to direct phototactic movement in motile species. *Haloquadratum*, the flat, square archaeon, lacks motility, and accordingly it possesses only bacteriorhodopsin and halorhodopsin. *Salinibacter*, phylogenetically unrelated to the archaea, possesses a light-driven retinal-containing proton pumps named xanthorhodopsin, a putative light-driven chloride pump, and retinal-based light sensors. Transcriptome sequencing showed that the genes coding for xanthorhodopsin are highly expressed (González-Torres et al., 2015). The salinixanthin carotenoid serves as a light antenna that transfers energy to the retinal group of xanthorhodopsin (Balashov et al., 2005; Balashov & Lanyi, 2007; Oren, 2013a). The retinal proteins of the halophilic prokaryotes are purple colored, with an absorbance maximum at 570–580 nm.

Few attempts have been made to quantitatively assess the presence of retinal proteins in red hypersaline brines. As these pigments are not extracted by organic solvents, their quantification is technically more difficult.

Retinal proteins were found in the microbial bloom in the Dead Sea in the summer of 1980 (1.9×10^7 prokaryotic cells/mL, up to 0.4 nmol bacteriorhodopsin-like protein per mg total protein) (Oren & Shilo, 1981). Bacteriorhodopsin was found at concentrations up to 2.2 mM in the crystallizer ponds of the Exportadora de Sal, Baja California, Mexico salterns (Javor, 1983) and in the saltern brines in Eilat (3.5×10^7 prokaryotes/mL; 3.6 nmol/L bacteriorhodopsin-like proteins) (Oren et al., 2016). In the salterns of Margherita di Savoia, Italy, bacteriorhodopsin of *Haloquadratum* (squarebop I) was found so abundantly that the possibility of extracting the compound from the concentrated biomass was explored for the isolation of functional bacteriorhodopsin (Corcelli, 2014; Lobasso et al., 2012). The potential presence of bacteriorhodopsin and halorhodopsin in other saltern brines was inferred from the analysis of environmental DNA extracted from biomass collected from the salterns at Santa Pola and Isla Cristina, Spain (Fernández et al., 2014), Sečovlje, Slovenia (Pašić et al., 2005), and Eilat, Israel (Ram-Mohan, Oren, & Papke, 2016).

4.3 What causes the red color of the brines?

There has been considerable confusion in the past about the question, what actually causes the color of the red brines. The color has been attributed to the presence of *Dunaliella* (e.g., Turpin, 1839), and even to the presence of the brine shrimp *Artemia salina*. Pierce (1914) wrote: "It is clear, then, that the small bacillus which gives to the concentrated brine in the salterns on the shores of the Bay of San Francisco their striking red color, may go over into the salt as it is harvested and stored." Baas-Becking (1931) also attributed the color to "pink, red and purple bacteria."

In the case of red blooms in the Dead Sea in 1980 and in 1992 there can be no doubt: orange β-carotene-rich strains of *Dunaliella* were never found in the lake, and all *Dunaliella* cells encountered were green. Therefore, the archaea with their bacterioruberin pigments were the cause of the red coloration (Oren, 1983b; Oren & Gurevich, 1993). The 1980 bloom also contained bacteriorhodopsin (Oren & Shilo, 1981), but no genes encoding retinal pigments were found in DNA extracted from the 1992 biomass (Bodaker, Suzuki, Oren, & Béjà, 2012).

Salinibacter makes up a minor part of the prokaryotic community in salterns, and HPLC studies never showed a dominance of salinixanthin over bacterioruberin pigments (Oren & Rodríguez-Valera, 2001). But the question whether orange *Dunaliella* cells or members of the *Halobacteria* are mainly responsible for the brine color shown in Fig. 1 is more difficult to answer.

When samples of the red saltern crystallizer brines from Eilat, from Newark, California, and from Mallorca, Spain were filtered, the filters were pink with the characteristic color of the bacterioruberin carotenoids (shown in supplementary fig. 6 in Viver et al., 2019). However, when those filters were extracted with organic solvents, the extracts were yellow-orange, and the absorption spectra were dominated by the 450 nm peak of β-carotene (Litchfield, Irby, Kis-Papo, & Oren, 2000; Oren & Dubinsky, 1994; supplementary fig. 6 in Viver et al., 2019). When biomass from such samples was collected by centrifugation, extraction of the pink pellet yielded a pink extract dominated by the 496 nm peak of α-bacterioruberin. The density of the β-carotene-rich *Dunaliella* cells is lower than that of the saturated crystallizer brine, and as a result, these cells are not precipitated during centrifugation. Thus, the filtration method provides a more reliable estimate of the total amounts of the different carotenoids present in the brine. Spectra of the extracts of filters always showed dominance of β-carotene, in spite of the fact that the optical properties of the surface of the filter are those of the *Halobacteria*. This paradox can be explained by the different distribution of the pigments in the microbial cells. In the archaea, the bacterioruberin carotenoids are distributed evenly in the cytoplasmic membrane (except for possible patches of "purple membrane" that contain the bacteriorhodopsin proton pump), while in *Dunaliella* the β-carotene is densely concentrated in granules attached to the thylakoid membranes in the chloroplast. As a result, the β-carotene has a very small in vivo optical cross-section because of the dense packing of the pigment (Oren, 2009a; Oren & Dubinsky, 1994). The conclusions of Pierce (1914) and Baas-Becking (1931) are thus supported by these analyses: the carotenoids of the red halophilic prokaryotes appear to be the main factor causing the characteristic red color of hypersaline brines.

The optical properties of the brines as caused by the different biota can be exploited for remote monitoring of saltern brines and other hypersaline environments. The remotely sensed infrared spectral information could be correlated with in situ field measurements. Remote sensing may thus be a useful approach to monitor salinity and population distributions of microbial communities (Dalton, Palmer-Moloney, Rogoff, Hlavka, & Duncan, 2009).

5. Natural hypersaline environments: Case studies
5.1 Great Salt Lake

In the end of the 19th century, Tilden (1898) described specific areas of the shore of Great Salt Lake being colored red, but there is no record of the extent of these areas and the nature of the color; it is well possible that those areas were partially evaporated hypersaline pools. Pink-pigmented microbes were first isolated from Great Salt Lake already in the first decades of the 20th century (Baxter, 2018; Daines, 1910; Frederick, 1924).

Today, red brines fill the entire northern part of Great Salt Lake, Utah. However, this is a relatively new phenomenon. After the completion of the railroad causeway in 1959–1960 that divides the fairly shallow (up to ~10 m deep) lake in two parts, the southern part that receives most of the freshwater inflow became less saline, while the salinity of the north arm increased to values >300 g/L, values approaching saturation. Thus, the proper conditions for the development of dense communities of red halophilic microorganisms were established. The microbiology of Great Salt Lake in the 1970s was well documented thanks to the studies by Fredrick Post. The north arm was populated by dense communities of *Dunaliella*, both the orange, β-carotene-rich *D. salina* and the green *D. viridis*, and of red archaea. The brine shrimp *Artemia salina* fed on the microbial communities (Post, 1977; Post, 1981). In the 1970s, *D. salina* was the dominant planktonic alga in the north arm (~332 g/L salt). Typical population densities of 200–1000 cells/L were reported, with peak values of 3000–10,000 cells/mL. Its horizontal distribution was highly patchy. In the north arm, *D. viridis* occurred mostly on the underside of rocks and wood along the shallow margin, out of direct sunlight. In the water column, *D. viridis* was reported to be more abundant in the deeper layers than at the surface. Thus, in August 1975, 4000 cells/mL were counted in samples from 4.5 m depth, an order of magnitude higher than the numbers in the upper 1.5 m (Oren, 2014a; Post, 1977).

Little qualitative and quantitative information was collected in the 1970s about the communities of the red archaea. Post identified the organisms as *Halobacterium* and *Halococcus*, which at the time were the only described genera within the group. He prepared an extensive collection of cultures for further study, but unfortunately the samples were not preserved. The list of prokaryotic strains isolated from Great Salt Lake and available from public culture collections given by Baxter and Zalar (2019) did not contain a single member of the *Halobacteria* class; *Halorhabdus utahensis*, isolated from a sediment sample collected from the south arm in the 1990s (Wainø, Tindall, & Ingvorsen, 2000) was not included in that list.

It is unfortunate that no further in-depth studies of the biota of Great Salt Lake were made for more than 20 years. Studies by local scientists were only resumed in the first years of this century. Most of these studies use molecular, cultivation-independent methods for the qualitative characterization of the communities. We also have a few quantitative data about the community densities. The total numbers of prokaryotes in the brines of the north arm collected in the period 2007–2010 was in the range of 4–10 × 10^7 cells/mL in the spring and summer months, decreasing to 1–3 × 10^6 in autumn and winter. Numbers of virus-like particles were two orders of magnitude higher (Baxter et al., 2011).

Analysis of 16S rRNA gene clone libraries of samples from different sites in the north arm and in the south arm collected in September and November 2004 yielded many novel phylotypes of *Halobacteria*. Genera represented were *Natronomonas, Halorhabdus, Halorubrum, Haloquadratum, Haloferax, Halogeometricum, Haloarcula,* and *Halobacterium*. In this study, *Salinibacter* was not detected following application of bacterial primers (Tazi, Breakwell, Harker, & Crandall, 2014). Molecular, 16S rRNA gene sequence-based evaluation of the prokaryotic communities in the less saline south arm (∼15% salt at the surface, ∼24.5% at 8 m depth) in June 2006 showed presence of organisms affiliated with the genera *Halogeometricum, Halonotius, Haloquadratum,* and *Halosimplex*. The relative abundance of *Halogeometricum* was inversely correlated with salinity, while the share of *Halonotius, Haloquadratum,* and *Halosimplex* increased with salinity. The study also included a sample from the north arm, but no results were reported (Meuser et al., 2013).

The most comprehensive cultivation-independent study of the biota of the north arm was based on samples collected at Rozel Point in the period 2003–2006. The water salinity varied between 24 and 30%. The majority of sequences retrieved were not closely related to any known species: only 64 out of 530 archaeal sequences showed >94% similarity with the recorded entries in the GenBank. The largest clade of clones was related to *Haloquadratum* (237 clones). This phylotype was present year-round. Other clades were related to *Halorubrum* (41 clones), *Natronococcus* (14 clones), and *Haloplanus* (4 clones). The remaining five clades did not show close similarity with any of the known groups. The study also included cultivation experiments. Up to 6.5 × 10^6 colony forming units per ml were obtained. Members of the genera *Halorubrum* and *Natronococcus* were detected using cultivation-independent approaches as well as by characterization of colonies on agar plates. *Salinibacter* was the most abundant member of the domain Bacteria, sequences retrieved being 97%–99% similar to *Salinibacter ruber* (Almeida-Dalmet, Sikaroodi, Gillivet, Litchfield, & Baxter, 2015).

5.2 The Dead Sea

Although red halophilic Archaea of the class *Halobacteria* are the major component of the biota of the Dead Sea, they seldom develop there to densities sufficient to impart a red color to the waters (Oren, 1988; Oren, 1997; Oren, 2000). Red halophilic prokaryotes were cultivated from the Dead Sea already in the 1930s—early 1940s (Volcani, 1944). The type strains of the species *Haloarcula marismortui*, *Haloferax volcanii*, *Halorubrum sodomense*, and *Halobaculum gomorrense* were all isolated from the lake.

The first quantitative microbiological studies in the Dead Sea, performed in 1963–1964, yielded numbers up to 8.9×10^6 prokaryotic cells per ml in the surface waters. Numbers decreased rapidly with depth. No red coloration of the waters was reported, but the dominance of red prokaryotes was obvious from the color of the pellets obtained by centrifugation of Dead Sea water. The dominant pigment was shown to be α-bacterioruberin, the main C_{50} carotenoid of the *Halobacteria* (Kaplan & Friedmann, 1970).

Although the total content of dissolved salts in the Dead Sea (~344 g/L) is not greatly different from that of red saltern brines, the dominance of divalent cations (currently ~2.04 M Mg^{2+} and 0.49 M Ca^{2+}) prevents massive development of blooms of the *Dunaliella* and red archaea. The pH of the water is slightly acidic (~6.0), lower than the values of 7–8 preferred by most members of the *Halobacteria* class. Development of red brines in the Dead Sea was reported only twice: in 1980 and in 1992. In both cases the growth of the microbial blooms was triggered by a significant dilution of the upper water layers by massive inflow of freshwater, following unusually rainy winters. The water added by the rain floods caused the formation of a stratified (meromictic) water column, interrupting the holomictic regime in which the water column (maximum depth ~300 m) was subjected to annual mixing.

Following dilution of the upper ~5 m of the Dead Sea water column with ~20% freshwater in the winter of 1979–1980, a bloom of green *Dunaliella* cells (*D. parva*) developed in the summer of 1980, reaching densities of up to 8800 cells/mL in the end of July. Availability of phosphate, the limiting inorganic nutrient in the Dead Sea, was another important factor regulating the extent of the bloom (Oren, 1985; Oren & Shilo, 1982). The input of organic material by *Dunaliella* as the primary producer enabled the development of red halophilic archaea up to 1.9×10^7 cells/mL in the upper meters of the water column in the summer of 1980, numbers sufficiently high to impart a red color to the water. The color remained visible for several weeks (Oren, 1983b, 1985). In addition to bacterioruberin carotenoids, presence of retinal proteins such as bacteriorhodopsin could be demonstrated in the biomass collected from the water (Oren & Shilo, 1981).

Meromixis ended in late 1982, and was followed by a holomictic episode that lasted until the end of 1991. Heavy rain floods during the winter of 1991–1992 initiated a new meromictic period: the upper 5 m of the water column became diluted to 70% of their normal salinity. Once more, massive blooms of *Dunaliella* and *Halobacteria* developed, reaching densities exceeding those monitored in 1980: numbers up to 1.5×10^4 *Dunaliella* cells/mL were observed, followed by development of up to 3×10^7 red archaea/mL. As a result, the waters of the lake were red once more, this time for several months (Oren, 1993, 2014a; Oren, Gurevich, Anati, et al., 1995; Oren, Gurevich, Gemmell, et al., 1995). Viruses attacking halophilic archaea may have been responsible for the subsequent decline in of the community: between 0.9 and 7.3×10^7 virus-like particles/mL were counted in the upper 20 m of the water column in October 1994 during the decline of the archaeal bloom (Oren et al., 1997).

Microscopical examination of the red prokaryotic community during the 1992 bloom mostly showed pleomorphic, flat cells, characteristic of many genera of *Halobacteria* such as *Haloferax* and *Haloarcula*. Analysis of the polar lipids found in biomass collected from the water during the bloom showed a simple pattern, dominated by the phytanyl diether lipid derivative of phosphatidylglycerol and the methyl ester of phosphatidylglycerol phosphate, found in all *Halobacteria*, and one major glycolipid, identical to the sulfated diglycosyl diether lipid (S-DGD-1) characteristic of *Haloferax* and a number of other genera. Glycolipids characteristic of genera such as *Halobacterium* or *Haloarcula* were not found (Oren & Gurevich, 1993). Samples of the material collected during the 1992 bloom were preserved, and this enabled later metagenomics analysis. Based on the 16S rRNA genes retrieved from the metagenome of the 1992 community, the bloom consisted of a single species of an organism remotely related to the genus *Halobacterium* of a type not yet brought into culture. By comparison, the community in 2007 remaining in the water column had a very low density, but showed a much higher diversity in the metagenomics analysis (Bodaker et al., 2010). No rhodopsin genes were found in the 1992 bloom, but novel bacteriorhodopsin and sensory rhodopsin genes were retrieved from Dead Sea water samples collected in 2007 and 2010 (Bodaker et al., 2012).

5.3 Alkaline hypersaline lakes

Great Salt lake and coastal saltern crystallizer ponds are examples of near-neutral environments that support dense communities of red microorganisms. Red brines are also found in highly alkaline hypersaline environments. Jannasch (1957) published beautiful pictures of red brines in the lakes of the Wadi an Natrun, Egypt, that have pH values around 11 and total salt concentrations exceeding 30%, dominated by sodium as the main cation and chloride, sulfate, and carbonate as the main anions. He attributed the coloration mainly to the presence of alkaliphilic photosynthetic sulfur bacteria. Later studies of the communities showed that bacterioruberin pigments of alkaliphilic members of the class *Halobacteria* are mainly responsible for the color of the brines (Imhoff, Sahl, Soliman, & Trüper, 1979; see further

Oren, 2013b). Similar communities of haloalkaliphilic members of the *Halobacteria* color Lake Magadi, Kenya and other East-African soda lakes red (Jones, Grant, Duckworth, & Owenson, 1998).

Brightly red brines are also seen in the concentrating ponds of salt production facilities fed with alkaline brines. Novel archaeal phylotypes of *Halobacteria* were retrieved from a crystallizer pond of an alkaline (pH 11.5–12) saltern at Lake Magadi, Kenya used for the production of common salt. Here the nanohaloarchaeal phylotype was first detected (Grant et al., 1999). Other systems of alkaline evaporation ponds are used for the production of soda ash (sodium carbonate). Exploration of the microbial diversity of the evaporation ponds of Botswana Ash (Pty) Ltd., Sua Pan, Sowa, Botswana, using cultivation-depentent as well as molecular-based culture-independent methods yielded isolates belonging to the genera *Natrialba*, *Natronococcus* and *Natronorubrum*, the environmental 16S rRNA gene library being dominated by the genera *Halorubrum*, *Natrialba*, *Natronorubrum* and a number of novel phylotypes (Gareeb & Setati, 2009). 16S rRNA gene libraries obtained from red microbial communities in soda ash concentration ponds of Abijata-Shalla Soda Ash, Ethiopia (pH 9.7–10.1) were dominated by *Halorubrum* and *Halorhabdus* phylotypes (Simachew et al., 2016).

6. Commercial salt production facilities

Multi-pond solar salterns, used worldwide for salt production in tropical and subtropical coastal areas, represent environments with increasing salt concentrations, from seawater to NaCl saturation. In most saltern systems, the crystallizer ponds develop the characteristic red color due to massive development of archaea of the *Halobacteria* class, *Dunaliella*, *Salinibacter*, and possibly additional types of pigmented microorganisms. As the salterns are generally easily accessible and operated in a reproducible way over the years, they have become a popular environment for the study of the adaptation of microbial communities to life at the highest salt concentrations. Development of carotenoid-rich *Dunaliella* cells in saltern crystallizer ponds was documented already in the 19th century (Dunal, 1838; Oren, 2005). It is somewhat surprising that so little is yet known about their in situ photosynthetic activity (Oren, 2009b). The communities of red prokaryotes in the salterns were already studied in the first decades of the 20th century (Baas-Becking, 1931; Hof, 1935; Pierce, 1914).

In-depth studies of the biology of salterns crystallizer brines were performed in many places worldwide. These studies included cultivation-dependent as well as cultivation-independent methods. Table 1 provides a non-exhaustive overview of studies on five continents that contributed

Table 1 Selected cultivation-dependent and cultivation-independent studies of the red microbial communities of saltern crystallizer ponds worldwide.

Geographical area, country	Location	Cultivation-dependent approaches	Cultivation-independent approaches	References
Europe				
Spain	Santa Pola, Alicante	*Haloferax, Haloarcula, Halorubrum, Halobacterium*		Rodriguez-Valera et al. (1985)
			16S rRNA gene amplification: phylotypes belonging to *Haloquadratum*	Benlloch, Martínez-Murcia, and Rodríguez-Valera (1996)
			16S rRNA gene amplification: phylotypes belonging to *Haloquadratum*	Benlloch, Acinas, Martínez-Murcia, and Rodríguez-Valera (1996)
			Microbial counts, activity measurements	Guixa-Boixareu et al. (1996)
			FISH counts, enumeration of *Salinibacter*	Antón et al. (2000)
			5S rRNA fingerprints	Casamayor, Calderón-Paz, and Pedrós-Alió (2000)
			Microbial counts, heterotrophic production (amino acids uptake)	Pedrós-Alió et al. (2000)
		Halorubrum, Haloarcula	16S rRNA gene amplification, RISA: *Haloquadratum* (SPhT phylotype), *Halorubrum, Natronobacterium*	Benlloch et al. (2001)

		16S rRNA gene amplification and DGGE: *Haloquadratum* (SPhT phylotype), *Halorubrum*, *Haloarcula*, *Halorhabdus*, *Natronococcus*, *Salinibacter*	Benlloch et al. (2002)
		16S rRNA gene amplification and DGGE; RISA; T-RFLP	Casamayor et al. (2002)
		16S rRNA gene amplification and DGGE; flow cytometry	Estrada, Henriksen, Gasol, Casamayor, and Pedrós-Alió (2004)
		16S rRNA gene amplification and DGGE; microscopic counts	Gasol et al. (2004)
		Metagenomics: *Haloquadratum*, *Salinibacter*; description of "*Candidatus* Haloredivivus"	Ghai et al. (2011)
		Metagenomics: *Haloquadratum*, *Salinibacter*; Sequences related to bacteriorhodopsins and halorhodopsins correlate with the abundance of *Haloquadratum*	Fernandez et al. (2013) and Fernández et al. (2014)
	Halorubrum, *Haloterrigena*, *Halolamina* *Haloplanus*, *Natronomonas*, *Halorientalis*		Viver et al. (2015)
		Metagenomics: *Haloquadratum*, *Salinibacter*	Moller and Liang (2017)
La Trinitat, Ebro Delta, Taragona		Microbial counts, activity measurements	Guixa-Boixareu et al. (1996)

Continued

Table 1 Selected cultivation-dependent and cultivation-independent studies of the red microbial communities of saltern crystallizer ponds worldwide.—cont'd

Geographical area, country	Location	Cultivation-dependent approaches	Cultivation-independent approaches	References
			5S rRNA fingerprints	Casamayor et al. (2000)
			Microbial counts, heterotrophic production (amino acids uptake)	Pedrós-Alió et al. (2000)
		Halorubrum, *Halolamina*, *Haloplanus*, *Haloferax*, *Natronomonas*, *Halomicrobium*, *Halobellus*		Viver et al. (2015)
	Isla Cristina		Metagenomics: *Halorubrum* dominates; Sequences related to bacteriorhodopsins and halorhodopsins correlate with the abundance of *Halorubrum*	Fernández et al. (2014)
			Metagenomics: *Halorubrum* dominates	Moller and Liang (2017)
	Mallorca		FISH counts, enumeration of *Salinibacter*	Antón et al. (2000)
		Halorubrum, *Haloterrigena*, *Halolamina*, *Haloplanus*, *Haloferax*, *Halovivax*, *Halomicrobium*, *Halobellus*		Viver et al. (2015)
			Metagenomics: *Haloquadratum*, *Salinibacter*	Viver et al. (2019)
	Ibiza		FISH counts, enumeration of *Salinibacter*	Antón et al. (2000)

	Fromentera		Viver et al. (2015)
		Halorubrum, Haloarcula, Haloterrigena, Natronomonas	
	Grand Canary	FISH counts, enumeration of *Salinibacter*	Antón et al. (2000)
	Janubio and Fuerteventura. Canary Islands		Viver et al. (2015)
		Halorubrum, Haloarcula, Halolamina, Haloplanus, Haloferax, Halonotius	
	Redonda and Penalva, Alto Vinalopó valley, Alicante (inland salterns)	16S rRNA gene clone library: *Haloarcula, Halorubrum, Haloquadratum, Halobacterium*	Zafrilla et al. (2010)
	Odiel marshlands	16S rRNA gene clone library and high-throughput 16S rRNA sequencing: *Salinibacter, Halorubrum, Haloquadratum*	Gómez-Villegas et al. (2018)
Italy	Saline di Tarquinia	Bacterial 16S rRNA gene amplification and DGGE: *Salinibacter*	Barghini, Silvi, Aquilanti, Marcelli, and Fenice (2014)
	Margherita di Savoia, Bari	*bop* (bacterioopsin) gene characterization	Corcelli (2014)
		Lipidomics for the identification and quantification of different types of *Halobacteria* and *Salinibacter*	Lopalco et al. (2011)
Slovenia	Sečovlje	16S rRNA gene clone library: *Halorubrum, Haloterrigena, Haloarcula, Haloquadratum*; *bop* (bacterioopsin) gene clone library: *Halorubrum, Haloquadratum, Haloarcula*	Pašić et al. (2005)

Continued

Table 1 Selected cultivation-dependent and cultivation-independent studies of the red microbial communities of saltern crystallizer ponds worldwide.—cont'd

Geographical area, country	Location	Cultivation-dependent approaches	Cultivation-independent approaches	References
		Haloferax, Haloarcula, Haloterrigena, Halorubrum, Natrinema	16S rRNA gene clone library: *Halorubrum, Haloterrigena, Haloarcula*	Pašić et al. (2007)
Croatia	Ston	*Haloferax, Haloarcula, Haloterrigena, Halobacterium*	16S rRNA gene clone library: *Halorubrum, Haloplanus*	Pašić et al. (2007)
Bulgaria	Pomorie and Burgas, Burgas Bay		16S rRNA gene clone library: *Halanaeroarchaeum, Halorubrum, Halonotius, Halobellus, Halovenus*	Kambourova, Tomova, Boyadzhieva, Radchenkova, and Vasileva-Tonkova (2016)
Greece	Messolonghi, western Greece	*Haloarcula, Halobacterium, Halococcus, Haloferax, Halogeometricum, Halorubrum, Haloterrigena*	Oligonucleotide microarray (PhyloChip): not further identified members of the *Halobacteria*	Tsiamis et al. (2008)
Middle East				
Turkey	Inland salterns of Kaldirim and Kayacik, central Anatolia	*Halobacterium, Haloarcula, Halorubrum*		Birbir et al. (2007)
	Inland salterns of Bingöl, Fadlum, Kemah, Tuzlagözü, eastern Anatolia	*Halobacterium, Haloarcula, Halorubrum, Salinibacter*	FISH, 16S rRNA gene clone libraries: *Haloquadratum, Haloarcula, Halorubrum, Halonotius*	Çinar and Mutlu (2016)
	Çamaltı		16S rRNA gene clone libraries, ARDRA (bacteria only): *Salinibacter*	Mutlu and Güven (2015)

Israel	Eilat		Amino acids incorporation and its inhibition by bile salts	Oren (1990a)
			Amino acids incorporation and its inhibition by different antibiotics	Oren (1990b)
			Thymidine incorporation and its inhibition by different antibiotics	Oren (1990c)
			Pigment and polar lipid analyses	Litchfield et al. (2000)
			16S rRNA amplicon length heterogeneity studies	Litchfield, Oren, Irby, Sikaroodi, and Gillivet (2009)
			Microscopic counts of prokaryotes and *Dunaliella*; effect of different substrates on community respiration	Oren (2016c)
			bop (bacterioopsin) gene libraries: *Halorubrum, Haloarcula, Halomicrobium, Halosimplex, Haloquadratum*	Ram-Mohan et al. (2016)
Asia				
India	Kelambakkam, Marakanam and Vedaranyam, Tamil Nadu	*Haloferax, Halorubrum, Haloarcula, Halobacterium, Halogeometricum, Salinibacter*	16S rRNA gene clone libraries: *Halogeometricum, Haloferax, Haloarcula, Natronococcus, Halococcus, Natronomonas, Haloquadratum*	Manikandan et al. (2009)
	Ribandar, Goa	*Halococcus, Halorubrum. Haloarcula, Haloferax*		Mani et al. (2012)
South Korea	Daecheon		16S rRNA gene clone libraries: *Halorubrum, Natronomonas*	Park, Kang, and Rhee (2006)

Continued

Table 1 Selected cultivation-dependent and cultivation-independent studies of the red microbial communities of saltern crystallizer ponds worldwide.—cont'd

Geographical area, country	Location	Cultivation-dependent approaches	Cultivation-independent approaches	References
Australia				
	Geelong, Victoria	*Halorubrum, Haloferax, Halonotius*	16S rRNA gene clone libraries: *Halorubrum, Halonotius, Haloquadratum*	Burns et al. (2004a)
	Dry Creek, South Australia; Bajool, Queensland; Lara, Victoria		16S rRNA gene clone libraries: *Haloquadratum, Halorubrum, Haloplanus, Halonotius, Natronomonas*	Oh et al. (2009)
Africa				
Tunisia	Sfax		16S rRNA gene clone libraries: *Haloquadratum, Halorubrum, Halorhabdus, Salinibacter*	Baati et al. (2008)
			Microscopic enumeration of *Dunaliella* and heterotrophic protists	Elloumi, Guermazi, et al. (2009)
			Microscopic enumeration of *Dunaliella* and heterotrophic protists	Elloumi, Carrias, Ayadi, Sime-Ngando, and Bouain (2009)
			16S rRNA gene clone libraries from sediment samples: *Halorubrum, Haloquadratum, Halonotius*	Baati, Guermazi, Gharsallah, Sghir, and Ammar (2010)
			FISH, FISH, 16S rRNA gene amplification and DGGE: *Haloquadratum, Halorubrum, Halonotius, Natronomonas, Halogeometricum, Salinibacter*	Boujelben, Gomariz, et al. (2012)

			Flow cytometry with cell sorting; 16S rRNA gene clone libraries: *Haloquadratum*, *Halorubrum*, *Salinibacter*	Trigui et al. (2011)
Kenya	Near Lake Magadi (alkaline salterns for production of common salt)		16S rRNA gene clone libraries: *Natronobacterium*, *Natronococcus*, *Halorubrum*; first detection of Nanohaloarchaeota sequences	Grant et al. (1999)
Ethiopia	Near Lake Abijata (alkaline ponds for the production of soda ash)		16S rRNA gene sequencing; *Halorubrum*, *Halorhabdus*	Simachew et al. (2016)
Botswana	Sua Pan (alkaline ponds for the production of soda ash)	*Natrialba*, *Natronococcus*, *Natronorubrum*	16S rRNA gene clone libraries: *Halorubrum*, *Natronorubrum*, *Natrialba*, *Natronolimnobius*	Gareeb and Setati (2009)
North America				
USA	San Francisco Bay		Pigment and polar lipid analyses	Litchfield et al. (2000)
			Amplicon Length Heterogeneity	Litchfield and Gillevet (2002)
			Remote sensing	Dalton et al. (2009)
			16S rRNA gene clone library and metagenomics: *Halonotius* and *Haloquadratum*	Kimbrel et al. (2018)
	Chula Vista, San Diego	*Halobacterium*, *Haloarcula*, *Halorubrum*	16S rRNA gene clone library: *Halobacterium*, *Haloarcula*, *Halorubrum*	Bidle et al. (2005)
			Metagenomics: *Halorubrum* and other members of the *Halobacteria*	Moller and Liang (2017)

Continued

Table 1 Selected cultivation-dependent and cultivation-independent studies of the red microbial communities of saltern crystallizer ponds worldwide.—cont'd

Geographical area, country	Location	Cultivation-dependent approaches	Cultivation-independent approaches	References
Mexico	Baja California		Cell counts and bacteriorhodopsin quantification	Javor (1983)
		Haloarcula, Halobacterium, Halococcus (the only genera of Halobacteria recognized at the time)		Javor (1984)
		Halorubrum, Haloarcula, Salinibacter		Sabet et al. (2009)
Caribbean and South America				
Puerto Rico	Cabo Rojo		Metagenomics: dominance of *Haloquadratum* and *Salinibacter*; in addition *Halorubrum, Haloplanus, Halococcus*	Couto-Rodríguez and Montalvo-Rodríguez (2019)
Jamaica	Yallahs	High recovery of archaea on media prepared with natural brine and addition of a *Halobacterium* cell extract		Wais (1988)
Argentina	Salitral Negro, Colorada Grande, Guatraché		FISH, 16S rRNA gene amplification and DGGE: *Haloquadratum,* "*Candidatus* Halorediviuus," *Salinibacter*	Di Meglio et al. (2016)

Peru	Maras (inland salterns, 3380 m above sea level)	*Haloarcula, Halobacterium, Halorubrum, Halogeometricum, Salinibacter*	FISH and 16S rRNA gene clone library: *Haloquadratum, Halobacterium*	Maturrano et al. (2006)
Chile	Lo Valdivia, Curicó	*Halorubrum, Halarcula, Haloterrigena, Halonotius, Halolamina, Haloferax, Haloplanus, Halogeometricum, Natronoarchaeum, Halobacterium, Halovivax*		Viver et al. (2015)

ARDRA = Amplified Ribosomal DNA Restriction Analysis; FISH = Fluorescence In Situ Hybridization; DGGE = Denaturing Gradient Gel Electrophoresis; RISA = Ribosomal Internal Spacer Analysis; T-RFLP = Terminal-Restriction Fragments Length Polymorphism.

information about the biology of the red saltern brines. I did not list the many additional studies in which one or more red colonies that developed on agar plates were characterized and described as new species. Many of the type strains of the species recognized today in the class *Halobacteria* were isolated from saltern evaporation ponds. Further results from many of the studies listed in Table 1 are highlighted in other sections of this review.

The table starts with studies performed in the Bras del Port salterns, Santa Pola, near Alicante on the Mediterranean coast of Spain. The extensive, and often interdisciplinary work performed at that site since from the 1980s onward has set the standards for the study of extremely halophilic microorganism in their natural environment. Many novel, state-of-the-art techniques were first applied in those salterns, and we owe much of our insight in the biology of red saltern brines to the research performed at the Santa Pola salterns for over three decades (Ventosa, Fernández, León, Sánchez-Porro, & Rodriguez-Valera, 2014).

7. The role of pigmented microorganisms in salt production

The managers of salterns for the production of common salt from seawater or other sources of salty waters generally consider the development of the red colors in the crystallizer brines as positive, and they realize the importance of biological phenomena in the process of salt production. Pigmented microorganisms absorb solar radiation, and thereby they increase the water temperature and the evaporation rate. The idea that pigmented halophilic microorganisms may increase evaporation rates was recently explored in a study in which carotenoid-rich bacteria (*Arthrobacter* sp., *Planococcus* sp.) were shown to stimulate by up to 20% the evaporation of saline industrial reject brines resulting from reverse osmosis or electrodialysis reversal (Silva-Castro et al., 2019). Some salt production companies operate monitoring programs for analyses of nutrients, standing crop and associated biological phenomena in the ponds; others may employ specialized microbiologists as consultants (Javor, 2002).

There are indications that the microorganisms may directly interact with the brine and influence the formation of the halite (NaCl) crystals during the evaporation process. A scanning electron microscopy study of the halite produced in the salterns of Berre in the south of France and comparative studies of halite produced in the laboratory in the presence of red halophilic archaea showed that salt crystallization may be induced and oriented by the growth of bacterial colonies. It was suggested that multiplication of the cells and the regulation of their internal salt concentration may oversaturate the closely neighboring environment and therefore induce halite crystallization (Castanier, Perthuisot, Matrat, & Morvan, 1999). In a recently study from Indonesia, addition of a not further characterized consortium of pigmented halophilic microorganisms to crystallizer pond brine of resulted in the production of whiter salt with a lower content of impurities (magnesium, calcium) than salt produced in a control pond. The salt crystals produced in the pond amended with the microorganisms were more compact and cubical, compared to the more fragile crystals that precipitated in the untreated pond. However, salt yield was not improved by the addition of the microbial consortium (Chasanah et al., 2020).

Still, the "missing link" between saltworks biology and solar salt yield and quality has never unequivocally been identified. In addition to cyanobacterial and algal polysaccharides, proteins released by halophilic microorganisms, including the red archaeal communities in the crystallizer ponds, may negatively affect salt quality (Oren, 2009c). Examination of the in situ concentrations and the possible role of organic osmotic compounds produced by microorganisms (e.g., glycerol formed by *Dunaliella*, glycine betaine made by halophilic cyanobacteria, and other compounds involved in the metabolism of different types of halophilic microorganisms) did not lead to the identification of one or more compounds that may significantly affect the salt production process (Oren, 2010).

8. Concluding remarks

The red brines as found in natural hypersaline lakes and in the crystallizer ponds of man-made salterns for the production of salt from seawater are fascinating environments. They are inhabited by a surprisingly high diversity of microorganisms adapted to life at the highest salt concentrations. Cultivation-dependent as well as cultivation-independent studies based on the sequencing of DNA extracted from the brines have contributed much insight into the biology of the systems, including the recognition of phylogenetic lineages of halophiles that were unknown until recently. For the microbial ecologist, the red brines are attractive environments for study because of the high community densities and the relatively simple structure of the ecosystem.

Also the pigments that may contribute to the coloration of the red brines are markedly diverse. The possible contribution of algae of the genus *Dunaliella* was recognized nearly two centuries ago, and Pierce (1914) and Baas-Becking (1931) correctly assessed that the red prokaryotes are mainly responsible for the characteristic color. Among those bacteria we find a great diversity of pigments; some (α-bacterioruberin and derivatives) are known for many decades, others (salinixanthin) were discovered and characterized only recently. The discovery of bacteriorhodopsin (Oesterhelt & Stoeckenius, 1971), followed by the exploration of the diverse retinal proteins found in halophilic archaea as well as halophilic bacteria (*Salinibacter*), shows that 40- and 50-carbon carotenoids are not the sole contributors to the pigmentation of the microbiota in hypersaline lakes and saltern ponds.

Biotechnological exploitation of some of the pigments produced by halophilic microorganisms has boosted the interest in the biology of extremely hypersaline environments: β-carotene is industrially produced from *Dunaliella* species (Ben-Amotz et al., 2009), and bacteriorhodopsin is finding many applications in optical appliances (Hampp, 2000).

When Paul and Mormile presented a case for the protection of hypersaline environments from a microbiological perspective, they also mentioned their esthetic appeal (Paul & Mormile, 2017). Some ecosystems with red brines are indeed visually attractive. One of those is *Spiral Jetty*, constructed by Robert Smithson in 1970 at Rozel Bay on the shore of the northern arm (Gunnison Bay) of Great Salt Lake, Utah. Explaining the history of the construction of *Spiral Jetty*, the artist wrote:

> My concern with salt lakes began with my work in 1968 on the Mono Lake Site-Nonsite in California. Later I read a book called Vanishing Trails of Atacama by William Rudolph which described salt lakes (salars) in Bolivia in all stages of desiccation, and filled with micro bacteria that give the water surface a red color.... Because of the remoteness of Bolivia and because Mono Lake lacked a reddish color, I decided to investigate the Great Salt Lake in Utah. ... I called the Utah Park Development and spoke to Ted Tuttle, who told me that water in the Great Salt Lake north of the Lucin Cutoff, which cuts the lake in two, was the color of tomato soup. That was enough of a reason to go out there and have a look.
>
> **(Smithson, 1972)**

In 2008 I had the pleasure of visiting Robert Smithson's *Spiral Jetty* at Rozel Bay (Fig. 4). The pink-to-orange brine flowing through the earthwork spiral is esthetically very satisfying.

Fig. 4 Robert Smithson's *Spiral Jetty*, Rozel Bay, Utah, photographed from the air in July 2004 (left) and from the shore in May 2008. *Photographs courtesy of Prof. B. K. Baxter, Westminster College, Salt Lake City. Reproduced with permission.*

Acknowledgments

I thank Prof. Bonnie K. Baxter, Westminster College, Salt Lake City, for supplying the pictures of Robert Smithson's *Spiral Jetty*, Rozel Bay, Utah shown in Fig. 4 and for helpful discussions.

References

Almeida-Dalmet, S., Sikaroodi, M., Gillivet, P. M., Litchfield, C. D., & Baxter, B. K. (2015). Temporal study of the microbial diversity of the north arm of Great Salt Lake, Utah, U.S. *Microorganisms, 3*, 310–326.

Andrei, A.-Ş., Banciu, H. L., & Oren, A. (2012). Living with salt: Metabolic diversity in archaea inhabiting saline ecosystems. *FEMS Microbiology Letters, 330*, 1–9.

Antón, J., Oren, A., Benlloch, S., Rodríguez-Valera, F., Amann, R., & Rosselló-Mora, R. (2002). *Salinibacter ruber* gen. nov., sp. nov., a novel extreme halophilic member of the *Bacteria* from saltern crystallizer ponds. *International Journal of Systematic and Evolutionary Microbiology, 52*, 485–491.

Antón, J., Llobet-Brossa, E., Rodríguez-Valera, F., & Amann, R. (1999). Fluorescence in situ hybridization analysis of the prokaryotic community inhabiting crystallizer ponds. *Environmental Microbiology, 1*, 517–523.

Antón, J., Peña, A., Santos, F., Martínez-Garcia, M., Schmitt-Kopplin, P., & Rosselló-Mora, R. (2008). Distribution, abundance and diversity of the extremely halophilic bacterium *Salinibacter ruber*. *Saline Systems, 4*, 15.

Antón, J., Rosselló-Mora, R., Rodríguez Valera, F., & Amann, R. (2000). Extremely halophilic Bacteria in crystallizer ponds from solar salterns. *Applied and Environmental Microbiology, 66*, 3052–3057.

Baas-Becking, L. G. M. (1931). Historical notes on salt and salt-manufacture. *Scientific Monthly, 32*, 434–446.

Baati, H., Guermazi, S., Amdouni, R., Gharsallah, N., Sghir, A., & Ammar, E. (2008). Prokaryotic diversity of a Tunisian multipond solar saltern. *Extremophiles, 12*, 505–518.

Baati, H., Guermazi, S., Gharsallah, N., Sghir, A., & Ammar, E. (2010). Novel prokaryotic diversity in sediments of Tunisian multipond solar saltern. *Resesarch in Microbiology, 161*, 573–582.

Balashov, S. P., Imasheva, E. S., Boichenko, V. A., Antón, J., Wang, J. M., & Lanyi, J. K. (2005). Xanthorhodopsin: A proton pump with a light-harvesting carotenoid antenna. *Science, 309,* 2061–2064.

Balashov, S. P., & Lanyi, J. K. (2007). Xanthorhodopsin: Proton pump with a carotenoid antenna. *Cellular and Molecular Life Sciences, 64,* 2323–2328.

Barghini, P., Silvi, S., Aquilanti, A., Marcelli, M., & Fenice, M. (2014). Bacteria from marine salterns as a model of microorganisms adapted to high environmental variations. *Journal of Environmental Protection Ecology, 15,* 897–906.

Baxter, B. K. (2018). Great Salt Lake microbiology: A historical perspective. *International Microbiology, 21,* 79–95.

Baxter, B. K., Mangalea, M. R., Willcox, S., Sabet, S., Nagoulat, M.-N., & Griffiths, J. D. (2011). Haloviruses of Great Salt Lake: A model for understanding viral diversity. In A. Ventosa, A. Oren, & Y. Ma (Eds.), *Halophiles and hypersaline environments* (pp. 173–190). Berlin: Springer-Verlag.

Baxter, B. K., & Zalar, P. (2019). The extremophiles of Great Salt Lake: Complex microbiology in a dynamic hypersaline ecosystem. In J. Seckbach & P. Rampelotto (Eds.), *Model ecosystems in extreme environments* (pp. 57–143). Elsevier. Chapter 4.

Ben-Amotz, A., Katz, A., & Avron, M. (1982). Accumulation of β-carotene in halotolerant algae: Purification and characterization of β-carotene-rich globules from *Dunaliella bardawil* (Chlorophyceae). *Journal of Phycology, 18,* 529–537.

Ben-Amotz, A., Lers, A., & Avron, M. (1988). Stereoisomers of β-carotene and phytoene in the alga *Dunaliella bardawil*. *Plant Physiology, 86,* 1286–1291.

Ben-Amotz, A., Polle, J. E. W., & Subba Rao, D. W. (Eds.), (2009). *The alga Dunaliella*. Enfield, NH: Science Publishers; CRC Press.

Benlloch, S., Acinas, S. C., Antón, J., López-López, A., Luz, S. P., & Rodríguez-Valera, F. (2001). Archaeal biodiversity in crystallizer ponds from a solar saltern: Culture versus PCR. *Microbial Ecology, 41,* 12–19.

Benlloch, S., Acinas, S. G., Martínez-Murcia, A. J., & Rodríguez-Valera, F. (1996). Description of prokaryotic biodiversity along the salinity gradient of a multipond solar saltern by direct PCR amplification of 16S rDNA. *Hydrobiologia, 329,* 19–31.

Benlloch, S., López-López, A., Casamayor, E. O., Øverås, L., Goddard, V., Daae, F. L., et al. (2002). Prokaryotic genetic diversity throughout the salinity gradient of a coastal solar saltern. *Environmental Microbiology, 4,* 349–360.

Benlloch, S., Martínez-Murcia, A. J., & Rodríguez-Valera, F. (1996). Sequencing of bacterial and archaeal 16S rRNA genes directly amplified from a hypersaline environment. *Systematic and Applied Microbiology, 18,* 574–581.

Bidle, K., Amadio, W., Oliveira, P., Paulish, T., Hicks, S., & Earnest, C. (2005). A phylogenetic analysis of haloarchaea found in a solar saltern. *Bios, 76,* 89–96.

Birbir, M., Calli, B., Mertoglu, B., Elevi Bardavid, R., Oren, A., Ogmen, M. N., et al. (2007). Extremely halophilic Archaea from Tuz Lake, Turkey, and the adjacent Kaldirim and Kayacik salterns. *World Journal of Microbiology and Biotechnology, 23,* 309–316.

Bodaker, I., Sharon, I., Suzuki, M. T., Feingersch, R., Shmoish, M., Andreishcheva, E., et al. (2010). Comparative community genomics in the Dead Sea: An increasingly extreme environment. *The ISME Journal, 4,* 399–407.

Bodaker, I., Suzuki, M. T., Oren, A., & Béjà, O. (2012). Dead Sea bacteriorhodopsin revisited. *Environmental Microbiology Reports, 4,* 617–621.

Bolhuis, H., Palm, P., Wende, A., Falb, M., Rampp, M., Rodriguez-Valera, F., et al. (2006). The genome of the square archaeon *Haloquadratum walsbyi*: Life at the limits of water activity. *BMC Genomics, 7,* 169.

Bolhuis, H., te Poele, E. M., & Rodriguez-Valera, F. (2004). Isolation and cultivation of Walsby's square archaeon. *Environmental Microbiology, 6,* 1287–1291.

Boujelben, I., Gomariz, M., Martínez-Garcia, M., Santos, F., Peña, A., López, C., et al. (2012). Spatial and seasonal prokaryotic community dynamics in ponds of increasing salinity of Sfax solar saltern in Tunisia. *Antonie van Leeuwenhoek, 101*, 845–857.

Boujelben, I., Yarza, P., Almansa, C., Villamor, J., Maalej, S., Antón, J., et al. (2012). Virioplankton community structure in Tunisian solar salterns. *Applied and Environmental Microbiology, 78*, 7429–7437.

Burns, D. G., Camakaris, H. M., Janssen, P. H., & Dyall-Smith, M. L. (2004a). Combined use of cultivation-dependent and cultivation-independent methods indicates that members of most haloarchaeal groups in an Australian crystallizer pond are cultivable. *Applied and Environmental Microbiology, 70*, 5258–5265.

Burns, D. G., Camakaris, H. M., Janssen, P. H., & Dyall-Smith, M. L. (2004b). Cultivation of Walsby's square haloarchaeon. *FEMS Microbiology Letters, 238*, 469–473.

Burns, D. G., Janssen, P. H., Itoh, T., Kamekura, M., Li, Z., Jensen, G., et al. (2007). *Haloquadratum walsbyi* gen. nov., sp. nov., the square haloarchaeon of Walsby, isolated from saltern crystallizers in Australia and Spain. *International Journal of Systematic and Evolutionary Microbiology, 57*, 387–392.

Butinar, L., Sonjak, S., Zalar, P., Plemenitaş, A., & Gunde-Cimerman, N. (2005). Melanized halophilic fungi are eukaryotic members of microbial communities in hypersaline waters of solar salterns. *Botanica Marina, 48*, 73–79.

Casamayor, E. O., Calderón-Paz, J. I., & Pedrós-Alió, C. (2000). 5S rRNA fingerprints of marine bacteria, halophilic archaea and natural prokaryotic assemblages along a salinity gradient. *FEMS Microbiology Ecology, 34*, 113–119.

Casamayor, E. O., Massana, R., Benlloch, S., Øverås, L., Díez, B., Goddard, V. J., et al. (2002). Changes in archaeal, bacterial and eukaryal assemblages along a salinity gradient by comparison of genetic fingerprinting methods in a multipond solar saltern. *Environmental Microbiology, 4*, 338–348.

Castanier, S., Perthuisot, J. P., Matrat, M., & Morvan, J. Y. (1999). The salt ooids of Berre salt works (Bouches du Rhône, France): The role of bacteria in salt crystallization. *Sedimentary Geology, 125*, 9–21.

Chasanah, E., Pratitis, A., Ambarwati, D., Fithriani, D., Susilowati, R., & Marihati (2020). Application of halophilic bacteria in traditional solar salt pond; A preliminary study. In *EMBRIO 2019. IOP conference series: Earth and environmental sciences 404*, 012035.

Çinar, S., & Mutlu, M. B. (2016). Comparative analysis of prokaryotic diversity in solar salterns in eastern Anatolia (Turkey). *Extremophiles, 20*, 589–601.

Clementino, M. M., Vieira, R. P., Cardoso, A. M., Nascimento, A. P., Silveira, C. B., Riva, T. C., et al. (2008). Prokaryotic diversity on one of the largest hypersaline coastal lagoons in the world. *Extremophiles, 12*, 595–604.

Corcelli, A. (2014). Biotechnological potential of solar salt works: Focus on Squarebop I bacteriorhodopsin. *EUSalt, 2014*, 9–12.

Corcelli, A., Lattanzio, V. M. T., Mascolo, G., Babudri, F., Oren, A., & Kates, M. (2004). Novel sulfonolipid in the extremely halophilic bacterium *Salinibacter ruber*. *Applied and Environmental Microbiology, 70*, 6678–6685.

Couto-Rodríguez, R. L., & Montalvo-Rodríguez, R. (2019). Temporal analysis of the microbial community from the crystallizer ponds in Cabo Rojo, Puerto Rico, using metagenomics. *Genes, 10*, 422.

Daines, L. L. (1910). *Physiological experiments on some algae of Great Salt Lake*. M.Sc. thesis. University of Utah.

Dalton, J. B., Palmer-Moloney, L. J., Rogoff, D., Hlavka, C., & Duncan, C. (2009). Remote monitoring of hypersaline environments in San Francisco Bay, CA, USA. *International Journal of Remote Sensing, 30*, 2933–2949.

Darwin, C. (1839). *Journal of researches into the geology and natural history of the various countries visited by H.M.S. Beagle, under the command of Captain Fitzroy, R.N. from 1832 to 1836*. London: Henry Colburn.

Di Meglio, L., Santos, F., Gomariz, M., Almansa, C., López, C., Antón, J., et al. (2016). Seasonal dynamics of extremely halophilic microbial communities in three Argentinian salterns. *FEMS Microbiology Ecology, 92* fiw184.
Diaz-Munoz, G., & Montalvo-Rodriguez, R. (2005). Halophilic black yeast *Hortaea werneckii* in the Cabo Rojo solar salterns: Its first record for this extreme environment in Puerto Rico. *Caribbean Journal of Science, 41,* 360–365.
Dunal, M. F. (1838). Extrait d'un mémoire de M. F. Dunal sur les algues qui colorent en rouge certaines eaux des marais salants méditerranéens. *Annales de la Science Naturelle Botanique Séries, 2*(9), 172–174.
Dyall-Smith, M. L., Pfeiffer, F., Klee, K., Palm, P., Gross, K., Schuster, S. C., et al. (2011). *Haloquadratum walsbyi*: Limited diversity in a global pond. *PLoS One, 6,* e20968.
Elevi Bardavid, R., Ionescu, D., Oren, A., Rainey, F. A., Hollen, B. J., Bagaley, D. R., et al. (2007). Selective enrichment, isolation and molecular detection of *Salinibacter* and related extremely halophilic *Bacteria* from hypersaline environments. *Hydrobiologia, 576,* 3–13.
Elevi Bardavid, R., Khristo, P., & Oren, A. (2008). Interrelationships between *Dunaliella* and halophilic prokaryotes in saltern crystallizer ponds. *Extremophiles, 12,* 5–14.
Elevi Bardavid, R., & Oren, A. (2008). Dihydroxyacetone metabolism in *Salinibacter ruber* and in *Haloquadratum walsbyi*. *Extremophiles, 12,* 125–131.
Elloumi, J., Carrias, J.-F., Ayadi, H., Sime-Ngando, T., & Bouain, A. (2009). Communities structure of the planktonic halophiles in the solar saltern of Sfax, Tunisia. *Estuarine, Coastal and Shelf Science, 81,* 19–26.
Elloumi, J., Guermazi, W., Ayadi, H., Bouain, A., & Aleya, L. (2009). Abundance and biomass of prokaryotic and eukaryotic microorganisms coupled with environmental factors in an arid multi-pond solar saltern (Sfax, Tunisia). *Journal of the Marine Biology Association of the United Kingdom, 89,* 243–253.
Encyclopédie ou Dictionnaire Raisonné des Sciences, des Arts et des Métiers par une société des gens de lettres, & Quatorzième, T. (1765). *A Neufchastel*. Chez Samuel Faulche & Compagnie.
Estrada, M., Henriksen, P., Gasol, J. M., Casamayor, E. O., & Pedrós-Alió, C. (2004). Diversity of planktonic photoautotrophic microorganisms along a salinity gradient as depicted by microscopy, flow cytometry, pigment analysis and DNA-based methods. *FEMS Microbiology Ecology, 49,* 281–293.
Fernandez, A. B., Ghai, R., Belen Martin-Cuadrado, A., Sanchez-Porro, C., Rodriguez-Valera, F., & Ventosa, A. (2013). Metagenome sequencing of prokaryotic microbiota from two hypersaline ponds of a marine saltern in Santa Pola, Spain. *Genome Announcements, 1,* e00933-13.
Fernández, A. B., Vera-Gargallo, B., Sánchez-Porro, C., Ghai, R., Papke, R. T., Rodriguez-Valera, F., et al. (2014). Comparison of prokaryotic community structure from Mediterranean and Atlantic saltern concentrator ponds by a metagenomic approach. *Frontiers of Microbiology, 5,* 196.
Filker, S., Gimmler, A., Dunthorn, M., Mahé, D., & Stoeck, T. (2015). Deep sequencing uncovers protistan plankton diversity in the Portuguese Ria Formosa solar saltern ponds. *Extremophiles, 19,* 283–295.
Frederick, E. (1924). *On the bacterial flora of Great Salt Lake and the viability of other microorganisms in Great Salt Lake water*. M.Sc. thesis. University of Utah.
Gareeb, A. P., & Setati, M. E. (2009). Assessment of alkaliphilic haloarchaeal diversity in Sua Pan evaporator ponds in Botswana. *African Journal of Biotechnology, 8,* 259–267.
Gasol, J. M., Casamayor, E. O., Joint, I., Garde, K., Gustavson, K., Benlloch, S., et al. (2004). Control of heterotrophic prokaryotic abundance and growth rate in hypersaline planktonic environments. *Aquatic Microbial Ecology, 34,* 193–206.
Ghai, R., Pašić, L., Fernández, A. B., Martin-Cuadrado, A.-B., Mizuno, C. M., McMahon, K. D., et al. (2011). New abundant microbial groups in aquatic hypersaline environments. *Scientific Reports, 1,* 135.

Gomariz, M., Martínez-Garcia, M., Santos, F., Rodriguez, F., Capella-Gutiérrez, S., Gabaldón, T., et al. (2015). From community approaches to single-cell genomics: The discovery of ubiquitous hyperhalophilic Bacteroidetes generalists. *ISME Journal*, 9, 16–31.

Gómez-Villegas, P., Vigara, J., & León, R. (2018). Characterization of the microbial population inhabiting a solar saltern pond of the Odiel marshlands (SW Spain). *Marine Drugs*, 16, 332.

González, M. A., Gómez, P. I., & Polle, J. E. W. (2009). Taxonomy and physiology of the genus *Dunaliella*. In A. Ben-Amotz, J. E. W. Polle, & D. W. S. Rao (Eds.), *The alga Dunaliella* (pp. 15–44). Enfield, NH: CRC Press; Science Publishers.

González-Torres, P., Pryszcz, L. P., Santos, F., Martínez-Garcia, M., Gabaldón, T., & Antón, J. (2015). Interactions between closely related bacterial strains are revealed by deep transcriptome sequencing. *Applied and Environmental Microbiology*, 81, 8445–8456.

Grant, S., Grant, W. D., Jones, B. E., Kato, C., & Li, L. (1999). Novel archaeal phylotypes from an East African alkaline saltern. *Extremophiles*, 3, 139–145.

Grant, W. D., & Tindall, B. J. (1986). The alkaline saline environment. In R. A. Herbert & G. A. Codd (Eds.), *Microbes in extreme environments* (pp. 25–54). London: Academic Press.

Guixa-Boixareu, N., Calderón-Paz, J. I., Heldal, M., Bratbak, G., & Pedrós-Alió, C. (1996). Viral lysis and bacterivory as prokaryotic loss factors along a salinity gradient. *Aquatic Microbial Ecology*, 11, 215–227.

Guixa-Boixereu, N., Lysnes, K., & Pedrós-Alió, C. (1999). Viral lysis and bacterivory during a phytoplankton bloom in a coastal water microcosm. *Applied and Environmental Microbiology*, 65, 1949–1958.

Gunde-Cimerman, N., Zalar, P., de Hoog, S., & Plemenitaş, A. (2000). Hypersaline waters in salterns—Natural ecological niches for halophilic black yeasts. *FEMS Microbiology Ecology*, 32, 235–240.

Gupta, R. S., Naushad, S., & Baker, S. (2015). Phylogenomic analyses and molecular signatures for the class *Halobacteria* and its two major clades: A proposal for division of the class *Halobacteria* into an emended order *Halobacteriales* and two new orders, *Haloferacales* ord. nov. and *Natrialbales* ord. nov., containing the novel families *Haloferacaceae* fam. nov. and *Natrialbaceae* fam. nov. *International Journal of Systematic and Evolutionary Mircrobiology*, 65, 1050–1069.

Gupta, R. S., Naushad, S., Fabros, R., & Adeolu, M. (2016). A phylogenomic reappraisal of family-level divisions within the class *Halobacteria*: Proposal to divide the order *Halobacteriales* into the families *Halobacteriaceae*, *Haloarculaceae* fam. nov., and *Halococcaceae* fam. nov., and the order *Haloferacales* into the families, *Haloferacaceae* and *Halorubraceae* fam. nov. *Antonie van Leeuwenhoek*, 109, 565–587.

Hampp, N. A. (2000). Bacteriorhodopsin: Mutating a biomaterial into an optoelectronic material. *Applied Microbiology and Biotechnology*, 53, 633–639.

Hartmann, R., Sickinger, H.-D., & Oesterhelt, D. (1980). Anaerobic growth of halobacteria. *Proceedings of the National Academy of Sciences of the United States of America*, 77, 3821–3825.

Hof, T. (1935). Investigations concerning bacterial life in strong brines. *Recueil des Travaux Botaniques Néerlandais*, 32, 92–173.

Hong, H. P., & Choi, J. K. (2015). Can the halophilic ciliate *Fabrea salina* be used as a biocontrol of microalgae blooms in solar salterns? *Ocean Science Journal*, 50, 529–536.

Imhoff, J. F., Sahl, H. S., Soliman, G. S. H., & Trüper, H. G. (1979). The Wadi Natrun: Chemical composition and microbial mass developments in alkaline brines of eutrophic desert lakes. *Geomicrobiology Journal*, 1, 219–234.

Jannasch, H. W. (1957). Die bakterielle Rotfärbung der Salzseen des Wadi Natrun (Ägypten). *Archiv für Hydrobiologie*, 53, 425–433.

Javor, B. J. (1983). Planktonic standing crop and nutrients in a saltern ecosystem. *Limnology and Oceanography*, 28, 153–159.

Javor, B. J. (1984). Growth potential of halophilic bacteria isolated from solar salt environments: Carbon sources and salt requirements. *Applied and Environmental Microbiology, 48,* 352–360.

Javor, B. J. (2002). Industrial microbiology of solar salt production. *Journal of Industrial Microbiology and Biotechnology, 28,* 42–47.

Jeffrey, S. W., & Egeland, E. S. (2009). Pigments of green pigments of green and red forms of *Dunaliella,* and related *Chlorophytes.* In A. Ben-Amotz, J. E. W. Polle, & D. W. S. Rao (Eds.), *The alga Dunaliella* (pp. 111–145). Enfield, NH: CRC Press; Science Publishers.

Jhin, S. H., & Park, J. S. (2018). A new halophilic heterolobosean flagellate, *Aurem hypersalina* gen. n. et sp. n., closely related to the *Pleurostomum-Tulamoeba* clade: Implications for adaptive radiation of halophilic eukaryotes. *Journal of Eukaryotic Microbiology, 66,* 221–231.

Jones, B. E., Grant, W. D., Duckworth, A. W., & Owenson, G. G. (1998). Microbial diversity of soda lakes. *Extremophiles, 2,* 191–200.

Kambourova, M., Tomova, I., Boyadzhieva, I., Radchenkova, N., & Vasileva-Tonkova, E. (2016). Unusually high prokaryotic diversity in crystallizer pond, Pomorie salterns, Bulgaria, revealed by phylogenetic analysis. *Archaea, 2016,* 7459679.

Kaplan, I. R., & Friedmann, A. (1970). Biological productivity in the Dead Sea. Part I. Microorganisms in the water column. *Israel Journal of Chemistry, 8,* 513–528.

Kelly, M., Norgård, S., & Liaaen-Jensen, S. (1970). Bacterial carotenoids. XXXI. C_{50} carotenoids 5. Carotenoids of *Halobacterium salinarium,* especially bacterioruberin. *Acta Chemica Scandinavica, 24,* 2169–2182.

Kimbrel, J. A., Ballor, N., Wu, Y.-W., David, M. M., Hazen, T. C., Simmons, B. A., et al. (2018). Microbial community structure and functional potential along a hypersalinity gradient. *Frontiers of Microbiology, 9,* 1492.

King, G. M. (2015). Carbon monoxide as a metabolic energy source for extremely halophilic microbes: Implications for microbial activity in Mars regolith. *Proceedings of the National Academy of Sciences of the United States of America, 112,* 4465–4470.

Kushwaha, S. C., Gochnauer, M. B., Kushner, D. J., & Kates, M. (1974). Pigments and isoprenoid compounds in extremely and moderately halophilic bacteria. *Canadian Journal of Microbiology, 20,* 241–245.

Kushwaha, S. C., Kramer, J. K. G., & Kates, M. (1975). Isolation and characterization of C_{50} carotenoid pigments and other polar isoprenoids from *Halobacterium cutirubrum. Biochimica et Biophysica Acta, 398,* 303–313.

Legault, B. A., Lopez-Lopez, A., Alba-Casado, J. C., Doolittle, W. F., Bolhuis, H., Rodriguez-Valera, F., et al. (2006). Environmental genomics of "*Haloquadratum walsbyi*" in a saltern crystallizer indicates a large pool of accessory genes in an otherwise coherent species. *BMC Genomics, 7,* 171.

Litchfield, C. D. (1991). Red—The magic color for solar salt production. In J.-C. Hocquet & R. Palme (Eds.), *Das Salz in der Rechts- und Handelsgeschichte* (pp. 403–412). Schwaz: Berenkamp.

Litchfield, C. D., & Gillevet, P. M. (2002). Microbial diversity and complexity in hypersaline environments: A preliminary assessment. *Journal of Industrial Microbiology and Biotechnology, 28,* 48–55.

Litchfield, C. D., Irby, A., Kis-Papo, T., & Oren, A. (2000). Comparisons of the polar lipid and pigment profiles of two solar salterns located in Newark, California, U.S.A., and Eilat, Israel. *Extremophiles, 4,* 259–265.

Litchfield, C. D., Oren, A., Irby, A., Sikaroodi, M., & Gillivet, P. M. (2009). Temporal and salinity impacts on the microbial diversity at the Eilat, Israel solar salt plant. *Global NEST Journal, 11,* 86–90.

Lobasso, S., Lopalco, P., Angelini, R., Pollice, A., Laera, G., Milano, F., et al. (2012). Isolation of Squarebop I bacteriorhodopsin from biomass of coastal salterns. *Protein Expression and Purification, 84,* 73–79.

Lopalco, P., Lobasso, S., Baronio, M., Angelini, R., & Corcelli, A. (2011). Impact of lipidomics on the microbial world of hypersaline environments. In A. Ventosa, A. Oren, & Y. Ma (Eds.), *Halophiles and hypersaline environments* (pp. 123–135). Berlin: Springer-Verlag.

Lutnæs, B. F., Oren, A., & Liaaen-Jensen, S. (2002). New C_{40}-carotenoid acyl glycoside as principal carotenoid of *Salinibacter ruber*, an extremely halophilic eubacterium. *Journal of Natural Products*, 65, 1340–1343.

Mancinelli, R. L., & Hochstein, L. I. (1986). The occurrence of denitrification in extremely halophilic bacteria. *FEMS Microbiology Letters*, 35, 55–58.

Mani, K., Salgaonkar, B. B., & Braganca, J. M. (2012). Culturable halophilic archaea at the initial and crystallization stages of salt production in a natural solar saltern of Goa, India. *Aquatic Biosystems*, 8, 15.

Manikandan, M., Kannan, V., & Pašić, L. (2009). Diversity of microorganisms in solar salterns of Tamil Nadu, India. *World Journal of Microbiology and Biotechnology*, 25, 1007–1017.

Maturrano, L., Santos, F., Rosselló-Mora, R., & Antón, J. (2006). Microbial diversity in Maras salterns, a hypersaline environment in the Peruvian Andes. *Applied and Environmental Microbiology*, 72, 3887–3895.

McGenity, T. J., & Oren, A. (2012). Hypersaline environments. In E. M. Bell (Ed.), *Life at extremes. Environments, organisms and strategies for survival* (pp. 402–437). UK: CABI International.

Meuser, J. E., Baxter, B. K., Spear, J. R., Peters, J. W., Posewitz, M. C., & Boyd, E. S. (2013). Contrasting patterns of community assembly in the stratified water column of Great Salt Lake, Utah. *Microbial Ecology*, 66, 268–280.

Mohan, N. R., Fullmer, M. S., Makkay, A. M., Wheeler, R., Ventosa, A., et al. (2014). Evidence from phylogenetic and genome fingerprinting analysis suggests rapidly changing variations in *Halorubrum* and *Haloarcula* populations. *Frontiers in Microbiology*, 5, 143.

Moller, A. G., & Liang, C. (2017). Determining virus-host interactions and glycerol metabolism profiles in geographically diverse solar salterns with metagenomics. *PeerJ*, 5, e2844.

Mongodin, E. F., Nelson, K. E., Daugherty, S., Deboy, R. T., Wister, J., Khouri, H., et al. (2005). The genome of *Salinibacter ruber*: Convergence and gene exchange among hyperhalophilic bacteria and archaea. *Proceedings of the National Academy of Sciences of the United States of America*, 103, 18147–18152.

Mora-Ruiz, M. D. R., Cifuentes, A., Font-Verdera, F., Pérez-Fernández, C., Farias, M. E., González, B., et al. (2018). Biogeographical patterns of bacterial and archaeal communities from distant hypersaline environments. *Systematic and Applied Microbiology*, 41, 139–150.

Mullakhanbhai, M. F., & Larsen, H. (1975). *Halobacterium volcanii* spec. nov., a Dead Sea halobacterium with a moderate salt requirement. *Archives of Microbiology*, 104, 207–214.

Munoz, R., Rosselló-Móra, R., & Amann, R. (2016). Revised phylogeny of *Bacteroidetes* and proposal of sixteen new taxa and two new combinations including *Rhodothermaeota* phyl. nov. *Systematic and Applied Microbiology*, 39, 281–296.

Mutlu, M. B., & Güven, K. (2015). Bacterial diversity in Çamaltı saltern, Turkey. *Polish Journal of Microbiology*, 64, 37–45.

Mutlu, M. B., Martínez-García, M., Santos, F., Peña, A., Guven, K., & Antón, J. (2008). Prokaryotic diversity in Tuz Lake, a hypersaline environment in inland Turkey. *FEMS Microbiology Ecology*, 65, 474–483.

Narasingarao, P., Podell, S., Ugalde, J. A., Brochier-Armanet, C., Emerson, J. B., Brocks, J. J., et al. (2011). De novo metagenomic assembly reveals abundant novel major lineage of Archaea in hypersaline microbial communities. *The ISME Journal*, 6, 81–93.

Oesterhelt, D., & Stoeckenius, W. (1971). Rhodopsin-like protein from the purple membrane of *Halobacterium halobium*. *Nature New Biology*, 233, 149–152.

Oesterhelt, D., & Stoeckenius, W. (1973). Functions of a new photoreceptor membrane. *Proceedings of the National Academy of Sciences of the United States of America, 70*, 2853–2857.

Oh, D., Porter, K., Russ, B., Burns, D., & Dyall-Smith, M. (2009). Diversity of *Haloquadratum* and other haloarchaea in three, geographically distant, Australian saltern crystallizer ponds. *Extremophiles, 14*, 161–169.

Oren, A. (1983a). *Halobacterium sodomense* sp. nov., a Dead Sea *Halobacterium* with an extremely high magnesium requirement. *International Journal of Systematic Bacteriology, 33*, 381–386.

Oren, A. (1983b). Population dynamics of halobacteria in the Dead Sea water column. *Limnology and Oceanography, 28*, 1094–1103.

Oren, A. (1985). The rise and decline of a bloom of halobacteria in the Dead Sea. *Limnology and Oceanography, 30*, 911–915.

Oren, A. (1988). The microbial ecology of the Dead Sea. In K. C. Marshall (Ed.), *Advances in microbial ecology: Vol. 10.* (pp. 193–229). New York: Plenum Publishing Company.

Oren, A. (1990a). Estimation of the contribution of halobacteria to the bacterial biomass and activity in a solar saltern by the use of bile salts. *FEMS Microbiology Ecology, 73*, 41–48.

Oren, A. (1990b). The use of protein synthesis inhibitors in the estimation of the contribution of halophilic archaebacteria to bacterial activity in hypersaline environments. *FEMS Microbiology Ecology, 73*, 187–192.

Oren, A. (1990c). Thymidine incorporation in saltern ponds of different salinities: Estimation of in situ growth rates of halophilic archaeobacteria and eubacteria. *Microbial Ecology, 19*, 43–51.

Oren, A. (1993). The Dead Sea—Alive again. *Experientia, 49*, 518–522.

Oren, A. (1995). Uptake and turnover of acetate in hypersaline environments. *FEMS Microbiology Ecology, 18*, 75–84.

Oren, A. (1997). Microbiological studies in the Dead Sea: 1892-1992. In T. Niemi, Z. Ben-Avraham, & J. R. Gat (Eds.), *The Dead Sea—The lake and its setting* (pp. 205–213). New York: Oxford University Press.

Oren, A. (2000). Biological processes in the Dead Sea as influenced by short-term and long-term salinity changes. *Archiv für Hydrobiologie, Special Issues Advances in Limnology, 55*, 531–542.

Oren, A. (2002). *Halophilic microorganisms and their environments.* Dordrecht: Kluwer Scientific Publishers.

Oren, A. (2005). A hundred years of *Dunaliella* research—1905-2005. *Saline Systems, 1*, 2.

Oren, A. (2008). Microbial life at high salt concentrations: Phylogenetic and metabolic diversity. *Saline Systems, 4*, 2.

Oren, A. (2009a). Saltern evaporation ponds as model systems for the study of primary production processes under hypersaline conditions. *Aquatic Microbial Ecology, 56*, 193–204.

Oren, A. (2009b). Microbial diversity and microbial abundance in salt-saturated brines: Why are the waters of hypersaline lakes red? In A. Oren, D. L. Naftz, P. Palacios, & W. A. Wurtsbaugh (Eds.), *Saline lakes around the world: Unique systems with unique values: Vol. 15.* (pp. 247–255). Utah State University: The S.J. and Jessie E. Quinney Natural Resources Research Library, College of Natural Resources; Natural Resources and Environmental Issues. Article 49.

Oren, A. (2009c). The microbiology of saltern crystallizer ponds and salt quality—A search for the "missing link" In S. Zuoliang (Ed.), *Proceedings of the 9th international symposium on salt: Vol. 2.* (pp. 904–912). Beijing: Gold Wall Press.

Oren, A. (2010). Thoughts on the "missing link" between saltworks biology and solar salt quality. *Global NEST Journal, 12*, 417–425.

Oren, A. (2011). Diversity of halophiles. In K. Horikoshi (Ed.), *Extremophiles handbook* (pp. 309–325). Tokyo: Springer.

Oren, A. (2013a). *Salinibacter*, an extremely halophilic bacterium with archaeal properties. *FEMS Microbiology Letters, 342*, 1–9.

Oren, A. (2013b). Two centuries of microbiological research in the Wadi Natrun, Egypt: A model system for the study of the ecology, physiology, and taxonomy of haloalkaliphilic microorganisms. In J. Seckbach, A. Oren, & H. Stan-Lotter (Eds.), *Polyextremophiles—Organisms living under multiple forms of stress* (pp. 103–119). Dordrecht: Springer.

Oren, A. (2013c). Life at high salt concentrations. In E. Rosenberg, E. F. DeLong, F. Thompson, S. Lory, & E. Stackebrandt (Eds.), *The Prokaryotes. A handbook on the biology of bacteria: Ecophysiology and biochemistry* (4th ed., pp. 421–440). New York: Springer.

Oren, A. (2014a). The ecology of *Dunaliella* in high-salt environments. *Journal of Biological Research Thessaloniki, 21*, 23.

Oren, A. (2014b). The family Halobacteriaceae. In E. Rosenberg, E. F. DeLong, F. Thompson, S. Lory, & E. Stackebrandt (Eds.), *The Prokaryotes. A handbook on the biology of bacteria: Ecophysiology and biochemistry* (4th ed., pp. 41–121). Berlin: Springer-Verlag.

Oren, A. (2015). Pyruvate: A key nutrient in hypersaline environments? *Microorganisms, 3*, 407–416.

Oren, A. (2016a). Life in hypersaline environments. In C. J. Hurst (Ed.), *Their world: A diversity of microbial environments. Advances in environmental microbiology: Vol. 1.* (pp. 301–339). Switzerland: Springer International Publishing.

Oren, A. (2016b). Life in high-salinity environments. In M. Yates, C. Nakatsu, R. Miller, & S. Pillai (Eds.), *Manual of environmental microbiology* (4th ed., pp. 4.3.2-1–4.3.2-13). Washington, DC: ASM Press.

Oren, A. (2016c). Probing saltern brines with an oxygen electrode: What can we learn about the community metabolism in hypersaline systems? *Life, 6*, 23.

Oren, A. (2017). Glycerol metabolism in hypersaline environments. *Environmental Microbiology, 19*, 851–863.

Oren, A. (2019a). Salinibacter, the red bacterial extreme halophile. In *Encyclopedia of life sciences*. Chichester, UK: John Wiley & Sons, Ltd. www.els.net

Oren, A. (2019b). Halophilic Archaea. In T. Schmidt (Ed.), *Encyclopedia of microbiology* (4th ed., pp. 495–503). Elsevier; Academic Press.

Oren, A., Abu-Ghosh, S., Argov, T., Kara-Ivanov, E., Shitrit, D., Volpert, A., et al. (2016). Expression and functioning of retinal-based proton pumps in a saltern crystallizer brine. *Extremophiles, 20*, 69–77.

Oren, A., Bratbak, G., & Heldal, M. (1997). Occurrence of virus-like particles in the Dead Sea. *Extremophiles, 1*, 143–149.

Oren, A., & Dubinsky, Z. (1994). On the red coloration of saltern crystallizer ponds. II. Additional evidence for the contribution of halobacterial pigments. *International Journal of Salt Lake Research, 3*, 9–13.

Oren, A., Duker, S., & Ritter, S. (1996). The polar lipid composition of Walsby's square bacterium. *FEMS Microbiology Letters, 138*, 135–140.

Oren, A., & Gurevich, P. (1993). Characterization of the dominant halophilic archaea in a bacterial bloom in the Dead Sea. *FEMS Microbiology Ecology, 12*, 249–256.

Oren, A., & Gurevich, P. (1994). Production of D-lactate, acetate, and pyruvate from glycerol in communities of halophilic archaea in the Dead Sea and in saltern crystallizer ponds. *FEMS Microbiology Ecology, 14*, 147–156.

Oren, A., Gurevich, P., Anati, D. A., Barkan, E., & Luz, B. (1995). A bloom of *Dunaliella parva* in the Dead Sea in 1992: Biological and biogeochemical aspects. *Hydrobiologia, 297*, 173–185.

Oren, A., Gurevich, P., Gemmell, R. T., & Teske, A. (1995). *Halobaculum gomorrense* gen. nov., sp. nov., a novel extremely halophilic archaeon from the Dead Sea. *International Journal of Systematic Bacteriology, 45*, 747–754.

Oren, A., & Meng, F.-W. (2019). 'Red—The magic color for solar salt production'—but since when? *FEMS Microbiology Letters*, *366*, fnz050.
Oren, A., Pri-El, N., Shapiro, O., & Siboni, N. (2006). Buoyancy studies in natural communities of square gas-vacuolate archaea in saltern crystallizer ponds. *Saline Systems*, *2*, 4.
Oren, A., & Rodríguez-Valera, F. (2001). The contribution of halophilic Bacteria to the red coloration of saltern crystallizer ponds. *FEMS Microbiology Ecology*, *36*, 123–130.
Oren, A., Rodríguez-Valera, F., Antón, J., Benlloch, S., Rosselló-Mora, R., Amann, R., et al. (2004). Red, extremely halophilic, but not archaeal: The physiology and ecology of *Salinibacter ruber*, a bacterium isolated from saltern crystallizer ponds. In A. Ventosa (Ed.), *Halophilic microorganisms* (pp. 63–76). Berlin: Springer-Verlag.
Oren, A., & Shilo, M. (1981). Bacteriorhodopsin in a bloom of halobacteria in the Dead Sea. *Archives of Microbiology*, *130*, 185–187.
Oren, A., & Shilo, M. (1982). Population dynamics of *Dunaliella parva* in the Dead Sea. *Limnology and Oceanography*, *27*, 201–211.
Øvreås, L., Daae, F. L., Torsvik, V., & Rodríguez-Valera, F. (2003). Characterization of microbial diversity in hypersaline environments by melting profiles and reassociation kinetics in combination with terminal restriction fragment length polymorphism (T-RFLP). *Microbial Ecology*, *46*, 291–301.
Park, J. S., Cho, B. C., & Simpson, A. G. B. (2006). *Halocafeteria seosinensis* gen. et sp. nov. (Bicosoecida), a halophilic bacteriovorous nanoflagellate isolated from a solar saltern. *Extremophiles*, *10*, 493–504.
Park, S.-J., Kang, C.-H., & Rhee, S.-K. (2006). Characterization of the microbial diversity in a Korean saltern by 16S rRNA gene analysis. *Journal of Microbiology and Biotechnology*, *16*, 1640–1645.
Park, J. S., Kim, H., Choi, D. H., & Cho, B. C. (2003). Active flagellates grazing on prokaryotes in high salinity waters of a solar saltern. *Aquatic Microbial Ecology*, *33*, 173–179.
Park, J. S., & Simpson, A. G. B. (2015). Diversity of heterotrophic protists from extremely hypersaline habitats. *Protist*, *166*, 422–437.
Pašić, L., Galán Bartual, S., Poklar Ulrih, N., Grabnar, M., & Herzog Velikonja, B. (2005). Diversity of halophilic archaea in the crystallizers of an Adriatic solar saltern. *FEMS Microbiology Ecology*, *54*, 491–498.
Pašić, L., Rodriguez-Mueller, B., Martin-Cuadrado, A.-B., Mira, A., Rohwer, F., & Rodriguez-Valera, F. (2009). Metagenomic islands of hyperhalophiles: The case of *Salinibacter ruber*. *BMC Genomics*, *10*, 570.
Pašić, L., Ulrih, N. P., Črnigoj, M., Grabnar, M., & Herzog Velikonja, B. (2007). Haloarchaeal communities in the crystallizers of two Adriatic solar salterns. *Canadian Journal of Microbiology*, *53*, 8–18.
Paul, V. G., & Mormile, M. R. (2017). A case for the protection of saline and hypersaline environments: A microbiological perspective. *FEMS Microbiology Ecology*, *93*, fix091.
Pedrós-Alió, C., Calderón-Paz, J. I., MacLean, M. H., Medina, G., Marrasé, C., Gasol, J. M., et al. (2000). The microbial food web along salinity gradients. *FEMS Microbiology Ecology*, *32*, 143–155.
Pierce, G. J. (1914). The behavior of certain micro-organisms in brine. *Carnegie Institution of Washington Publication no 193*, 49–69.
Podell, S., Emerson, J. B., Jones, C. M., Ugalde, J. A., Welch, S., Heidelberg, K. B., et al. (2014). Seasonal fluctuations in ionic concentrations drive microbial succession in a hypersaline lake community. *The ISME Journal*, *8*, 979–990.
Podell, S., Ugalde, J. A., Narasingarao, P., Banfield, J. F., & Heidelberg, K. B. (2013). Assembly-driven community genomics of a hypersaline microbial ecosystem. *PLoS One*, *8*, e61692.
Polle, J. E. W., Jin, E. S., & Ben-Amotz, A. (2020). The alga *Dunaliella* revisited: Looking back and moving forward with model and production organisms. *Algal Research*, *49*101948.

Polle, J. E. W., Roth, R., Ben-Amotz, A., & Goodenough, U. (2020). Ultrastructure of the green alga *Dunaliella salina* strain CCAP19/18 (Chlorophyta) as investigated by quick-freeze deep-etch electron microscopy. *Algal Research, 49*, 101953.

Post, F. J. (1977). The microbial ecology of the Great Salt Lake. *Microbial Ecology, 3*, 143–165.

Post, F. J. (1981). Microbiology of the Great Salt Lake north arm. *Hydrobiologia, 81*, 59–69.

Ram-Mohan, N., Oren, A., & Papke, R. T. (2016). Analysis of the bacteriorhodopsin-producing haloarchaea reveals a core community that is stable over time in the salt crystallizers of Eilat, Israel. *Extremophiles, 20*, 747–757.

Rodriguez-Valera, F., Ventosa, A., Juez, G., & Imhoff, J. F. (1985). Variation of environmental features and microbial populations with salt concentrations in a multi-pond saltern. *Microbial Ecology, 11*, 107–115.

Rosselló-Mora, R., Lee, N., Antón, J., & Wagner, M. (2003). Substrate uptake in extremely halophilic microbial communities revealed by microautoradiography and fluorescence in situ hybridization. *Extremophiles, 7*, 409–413.

Rosselló-Mora, R., Lucio, M., Peña, A., Brito-Echeverria, J., López-López, A., Calens-Vadell, M., et al. (2008). Metabolic evidence for biogeographic isolation of the extremophilic bacterium *Salinibacter ruber*. *The ISME Journal, 2*, 242–253.

Sabet, S., Diallo, L., Hays, L., Jung, W., & Dillon, J. G. (2009). Characterization of halophiles isolated from solar salterns in Baja California, Mexico. *Extremophiles, 13*, 643–656.

Santos, F., Yarza, P., Parro, V., Meseguer, I., Rosselló-Móra, R., & Antón, J. (2012). Culture-independent approaches for studying viruses from hypersaline environments. *Applied and Environmental Microbiology, 78*, 1635–1643.

Sher, J., Elevi, R., Mana, L., & Oren, A. (2004). Glycerol metabolism in the extremely halophilic bacterium *Salinibacter ruber*. *FEMS Microbiology Letters, 232*, 211–215.

Li Shizhen (1578). *Ben Cao Gang Mu* (Compendium of Materia Medica). 书名:本草纲目作者:(明) 李时珍 出版发行:上海:商务印书馆，1930 载体形态:6册 所引用部分:第三册；金石部；卷十一；食盐；38页.

Silva-Castro, G. A., Conrad Moyo, A., Khumalo, L., van Zyl, L. J., Petrik, L. F., & Trindade, M. (2019). Factors influencing pigment production by halophilic bacteria and its effect on brine evaporation rates. *Microbiology and Biotechnology, 12*, 334–345.

Simachew, A., Lanzen, A., Gessesse, A., & Øvreås, L. (2016). Prokaryotic community diversity along an increasing salt gradient in a soda ash concentration pond. *Microbial Ecology, 71*, 326–338.

Smithson, R. (1972). The Spiral Jetty. In G. Kepes (Ed.), *Arts of the environment*. New York: G. Braziller.

Sorokin, D. Y., Kublanov, I. V., Yakimov, M. M., Rijpstra, W. I., & Sinninghe Damsté, J. S. (2016). *Halanaeroarchaeum sulfurireducens* gen. nov., sp. nov., the first obligately anaerobic sulfur-respiring haloarchaeon, isolated from a hypersaline lake. *International Journal of Systematic and Evolutionary Microbiology, 66*, 2377–2381.

Sorokin, D. Y., Messina, E., Smedile, F., Roman, P., Sinninghe Damsté, J. S., Ciordia, S., et al. (2017). Discovery of anaerobic lithoheterotrophic haloarchaea, ubiquitous in hypersaline habitats. *The ISME Journal, 11*, 1245–1260.

Tazi, L., Breakwell, D. P., Harker, A. R., & Crandall, K. A. (2014). Life in extreme environments: Microbial diversity in Great Salt Lake, Utah. *Extremophiles, 18*, 525–535.

Teodoresco, E. C. (1905). Organisation et développement du *Dunaliella*, nouveau genre de Volvocacée-Polyblepharidée. *Beihefte zum botanischen Centralblatt, XVIII*, 215–232.

Tilden, J. E. (1898). American algae. In *Century III (Minneapolis, Minnesota)* (p. 298).

Tomlinson, G. A., & Hochstein, L. I. (1972). Studies on acid production during carbohydrate metabolism by extremely halophilic bacteria. *Canadian Journal of Microbiology, 18*, 1973–1976.

Torreblanca, M., Rodriguez-Valera, F., Juez, G., Ventosa, A., Kamekura, M., & Kates, M. (1986). Classification of non-alkaliphilic halobacteria based on numerical taxonomy and polar lipid composition, and description of *Haloarcula* gen. nov. and *Haloferax* gen. nov. *Systematic and Applied Microbiology, 8*, 89–99.

Trigui, H., Masmoudi, S., Brochier-Armanet, C., Barani, A., Grégori, G., Denis, M., et al. (2011). Characterization of heterotrophic prokaryote subgroups in the Sfax coastal solar salterns by combining flow cytometry cell sorting and phylogenetic analysis. *Extremophiles*, 15, 347–358.

Tsiamis, G., Katsaveli, K., Ntougias, S., Kyrpides, N., Andersen, G., Piceno, Y., et al. (2008). Prokaryotic community profiles at different operational stages of a Greek solar saltern. *Research in Microbiology*, 159, 609–627.

Turpin, P. J. F. (1839). Queleques observations nouvelles sur les Protococcus qui colorent en rouge les eaux des marais salants. *Comptes Rendus Hebdomadaires des Séances de l'Académie des Sciences – Sciences Naturelles*, 1839, 626–635.

Ventosa, A., de la Haba, R. R., Sánchez-Porro, C., & Papke, R. T. (2015). Microbial diversity of hypersaline environments: A metagenomic approach. *Current Opinion in Microbiology*, 25, 80–87.

Ventosa, A., Fernández, A. B., León, M. J., Sánchez-Porro, C., & Rodriguez-Valera, F. (2014). The Santa Pola saltern as a model for studying the microbial ecology of hypersaline environments. *Extremophiles*, 18, 811–824.

Viver, T., Cifuentes, A., Díaz, S., Rodríguez-Valdecantos, G., González, B., Antón, J., et al. (2015). Diversity of extremely halophilic cultivable prokaryotes in Mediterranean, Atlantic and Pacific solar salterns: Evidence that unexplored sites constitute sources of cultivable novelty. *Systematic and Applied Microbiology*, 38, 266–275.

Viver, T., Orellana, L. H., Díaz, S., Urdiain, M., Ramos-Barbero, M. D., González-Pastor, J. E., et al. (2019). Predominance of deterministic microbial community dynamics in salterns exposed to different light intensities. *Environmental Microbiology*, 21, 4300–4315.

Viver, T., Orellana, L., González-Torres, P., Díaz, S., Urdiain, M., Farías, M. E., et al. (2018). Genomic comparison between members of the *Salinibacteraceae* family, and description of a new species of *Salinibacter* (*Salinibacter altiplanensis* sp. nov.) isolated from high altitude hypersaline environments of the Argentinian Altiplano. *Systematic and Applied Microbiolology*, 41, 198–212.

Volcani, B. E. (1944). The microorganisms of the Dead Sea. In *Papers collected to commemorate the 70th anniversary of Dr. Chaim Weizmann* (pp. 71–85), Rehovoth: Daniel Sieff Research Institute. Collective volume.

Vreeland, R. H., Straight, S., Krammes, J., Dougherty, K., Rosenzweig, W. D., & Kamekura, M. (2002). *Halosimplex carlsbadense* gen. nov., sp. nov., a unique halophilic archaeon, with three 16S rRNA genes, that grows only in defined medium with glycerol and acetate or pyruvate. *Extremophiles*, 6, 445–452.

Wainø, M., Tindall, B. J., & Ingvorsen, K. (2000). *Halorhabdus utahensis* gen. nov., sp. nov., an aerobic, extremely halophilic member of the *Archaea* from Great Salt Lake, Utah. *International Journal of Systematic and Evolutionary Microbiology*, 50, 183–190.

Wais, A. C. (1988). Recovery of halophilic archaebacteria from natural environments. *FEMS Microbiology Ecology*, 4, 211–216.

Walsby, A. E. (1980). A square bacterium. *Nature*, 283, 69–71.

Youssef, N. H., Savage-Ashlock, K. N., McCully, A. L., Luedtke, B., Shaw, E. I., et al. (2014). Trehalose/2-sulfotrehalose biosynthesis and glycine betaine uptake are widely spread mechanisms for osmoadaptation in the *Halobacteriales*. *The ISME Journal*, 8, 636–649.

Zafrilla, B., Martínez-Espinosa, R. M., Alonso, M. A., & Bonete, M. J. (2010). Biodiversity of Archaea and floral of two inland saltern ecosystems in the Alto Vinalopó Valley, Spain. *Saline Systems*, 6, 10.

CHAPTER THREE

Clostridium thermocellum: A microbial platform for high-value chemical production from lignocellulose

R. Mazzoli[a,]* and D.G. Olson[b,c]
[a]Structural and Functional Biochemistry, Laboratory of Proteomics and Metabolic Engineering of Prokaryotes, Department of Life Sciences and Systems Biology, University of Torino, Torino, Italy
[b]Thayer School of Engineering, Dartmouth College, Hanover, NH, United States
[c]Center for BioEnergy Innovation, Oak Ridge National Laboratory, Oak Ridge, TN, United States
*Corresponding author: e-mail address: roberto.mazzoli@unito.it

Contents

1. Introduction	113
2. Overview of *C. thermocellum* central metabolism	116
3. Improving fermentation of hemicellulose-derived sugars by *C. thermocellum*	120
4. Improving production of high-value chemicals in *C. thermocellum* by metabolic engineering	122
4.1 Ethanol	122
4.2 *n*-Butanol and isobutanol	135
4.3 Lactic acid	142
4.4 Medium-chain esters	146
5. Conclusion	147
Acknowledgments	150
Conflict of interest	150
References	150

Abstract

Second generation biorefining, namely fermentation processes based on lignocellulosic feedstocks, has attracted tremendous interest (owing to the large availability and low cost of this biomass) as a strategy to produce biofuels and commodity chemicals that is an alternative to oil refining. However, the innate recalcitrance of lignocellulose has slowed progress toward economically viable processes. Consolidated bioprocessing (CBP), i.e., single-step fermentation of lignocellulose may dramatically reduce the current costs of 2nd generation biorefining. Metabolic engineering has been used as a tool to develop improved microbial strains supporting CBP. *Clostridium thermocellum* is among the most efficient cellulose degraders isolated so far and one of the most promising host organisms for application of CBP. The development of efficient and reliable genetic tools has allowed significant progress in metabolic engineering of this strain aimed at expanding the panel of growth substrates and improving the production of a number of commodity chemicals of industrial interest such as ethanol, butanol, isobutanol, isobutyl acetate and lactic acid. The present review aims to summarize recent developments in metabolic engineering of this organism which currently represents a reference model for the development of biocatalysts for 2nd generation biorefining.

Abbreviations

2PG	2-phosphoglycerate
3PG	3-phosphoglycerate
Aat	alcohol acetyltransferase
Acetyl-P	acetyl phosphate
Ack	acetate kinase
Adh	alcohol dehydrogenase
AdhE	bifunctional alcohol/aldehyde dehydrogenase
Aldh	aldehyde dehydrogenase
Als	α-acetolactate synthase
Bcd/EftAB	butyryl-CoA dehydrogenase/electron transfer protein
BPG	1,3-bisphosphoglycerate
Cat	chloramphenicol acetyltransferase
CimA	citramalate synthase
Crt	crotonase
Dhad	dihydroxy acid dehydratase
DHAP	dihydroxy acetone phosphate
Eno	enolase
F6P	fructose 6-phosphate
Fba	fructose 1,6-bisphosphate aldolase
FBP	fructose 1,6-bisphosphate
Fd	ferredoxin
FNOR	ferredoxin:NAD oxidoreductase
G6P	glucose 6-phosphate
GAP	glyceraldehyde 3-phosphate
Gapdh	glyceraldehyde 3-phosphate dehydrogenase
Glk	glucokinase
Gpi	glucose-6-phosphate isomerase
H_2ase	hydrogenase
Hbd	3-hydroxybutyryl-CoA dehydrogenase
Kari	keto acid reductoisomerase
Kdc	2-ketoacid decarboxylase
KivD	2-ketoisovalerate decarboxylase
Kor	ketoisovalerate ferredoxin-dependent reductase
IlvA	threonine dehydratase
Ldh	lactate dehydrogenase
LeuA	2-isopropylmalate synthase

LeuB	3-isopropylmalate dehydrogenase
LeuCD	isopropylmalate isomerase
MA	malic acid
Mdh	malate dehydrogenase
Me	malic enzyme
Nfn	NADH-dependent reduced ferredoxin:NADP$^+$ oxidoreductase
OAA	oxaloacetic acid
Pdc	pyruvate decarboxylase
PEP	phosphoenolpyruvate
Pepck	phosphoenolpyruvate carboxykinase
Pfk	phosphofructokinase
Pfl	pyruvate-formate lyase
Pfor	pyruvate ferredoxin/flavodoxin oxidoreductase
Pgk	phosphoglycerate kinase
Pgm	phosphoglycerate mutase
PP$_i$	pyrophosphate
Ppdk	pyruvate phosphate dikinase
Pta	phosphotransacetylase
Pyk	pyruvate kinase
Rex	global redox-responsive transcription factor
Rnf NAD$^+$	ion-translocating reduced ferredoxin oxidoreductase
Ter	trans-enoyl-CoA reductase
Thl	thiolase
Tpi	triosephosphate isomerase
Xi	xylose isomerase
Xk	xylulose kinase

1. Introduction

Second generation biorefining, namely industrial fermentation based on lignocellulosic feedstocks, has attracted substantial research interest over the last decades because of its potential as a sustainable alternative to oil refining to produce fuels and high-value chemicals. The large availability and low cost of this biomass make it an ideal feedstock for these processes. However, the innate recalcitrance of lignocellulose to biodegradation has been a significant barrier to the development of economically viable 2nd generation processes (Lynd, 2017). Traditionally, fermentation of lignocellulosic biomass has been carried out through multiple-step process configurations generally requiring biomass pretreatment by physical and/or chemical technologies followed by separate hydrolysis and fermentation of soluble carbohydrates (Lynd, Weimer, van Zyl, & Pretorius, 2002). A number of research groups have been working at the development of consolidated bioprocessing (CBP, that is single-step fermentation), based on its potential to dramatically reduce the cost of lignocellulose fermentation (Lynd, Van Zyl, McBride, & Laser, 2005). Metabolic engineering has been used to develop strains having both efficient plant-biomass degradation properties and the ability to produce industrially relevant compounds, with much effort focused on the bacterium *Clostridium thermocellum*.

C. thermocellum was among the first isolated cellulolytic microorganisms (McBee, 1954) (for more detailed description of isolation and early studies on characterization of *C. thermocellum* please refer to Akinosho, Yee, Close, & Ragauskas, 2014) and the first microbial species in which cellulosomes, namely efficient protein complexes for deconstruction of plant biomass, were discovered (Bayer, Kenig, & Lamed, 1983; Lamed, Setter, & Bayer, 1983). *C. thermocellum* is an anaerobic, Gram positive thermophile (optimal growth temperature 60 °C) among the best cellulose solubilizers isolated so far (Demain, Newcomb, & Wu, 2005; Lynd et al., 2002). It can consume cellulose at a high rate $2.5\,g/L\,h^{-1}$ (Argyros et al., 2011) which is more than an order of magnitude higher than that reported for other cellulolytic microorganisms, including *Clostridium cellulovorans*, *Clostridium cellulolyticum* (now *Ruminiclostridium thermocellum*) and *Caldicellulosiruptor bescii* (Desvaux, Guedon, & Petitdemange, 2000; Hamilton-Brehm et al., 2010; Sleat, Mah, & Robinson, 1984). Recent studies have found that *C. thermocellum* can solubilize woody and herbaceous lignocellulosic feedstocks with two- to fourfold higher efficiency than cellulases from fungi which are traditionally used for biomass saccharification (Holwerda et al., 2019; Lynd et al., 2016). Furthermore, the high temperatures required for growth of *C. thermocellum* reduce the risk of contamination (Akinosho et al., 2014). However, *C. thermocellum* cannot grow on hemicellulose and pectin (Aburaya, Esaka, Morisaka, Kuroda, & Ueda, 2015; Demain et al., 2005) which are important components (20%–25%) of the plant biomass (Gray, Zhao, & Emptage, 2006). The latter represents an issue affecting the use of *C. thermocellum* in CBP of lignocellulosic biomass and metabolic engineering strategies addressing it will be described in a dedicated section of this review.

Several *C. thermocellum* strains have been described so far, which include the type strain ATCC 27405, YS, LQRI, JW20, BC1, and DSM 1313 (Akinosho et al., 2014), but most research aimed at engineering biocatalysts for CBP of lignocellulosic biomass has involved strain DSM 1313, because of its higher genetic tractability (Akinosho et al., 2014), likely due to a relative lack of restriction-methylation systems, compared to other strains (Riley, Ji, Schmitz, Westpheling, & Guss, 2019). The complete sequence of the *C. thermocellum* DSM 1313 genome (Feinberg et al., 2011) and the first study reporting a method for targeted chromosomal gene modification in this strain (Tripathi et al., 2010) date back to about 10 years ago. Since then, impressive progress on metabolic engineering of this strain has been attained based on efficient and reliable methods for transformation and markerless gene deletion (Olson & Lynd, 2012). However, the type strain *C. thermocellum* ATCC 27405, which also received significant attention, is much more recalcitrant to gene modifications and a reliable protocol for its transformation was developed only in 2019 (Riley et al., 2019). As a result, all of the metabolic engineering studies reported in the next sections involve *C. thermocellum* DSM 1313. Genetic tools for modifying *C. thermocellum* DSM 1313 currently include a panel of autologous and heterologous transcriptional promoters (Olson et al., 2015) and inducible gene expression systems (Mearls, Olson, Herring, & Lynd, 2015). A Clustered Regularly-Interspaced Short Palindromic Repeat (CRISPR)/cas (CRISPR associated) system has been recently developed as an additional tool for editing the *C. thermocellum* genome (Walker et al., 2020). Furthermore, systems for fine tuning gene expression in *C. thermocellum* based on either riboswitches (Marcano-Velazquez, Lo, Nag, Maness, & Chou, 2019) or CRISPR interference (Ganguly, Martin-Pascual, & van Kranenburg, 2020) have been recently reported (Ganguly et al., 2020; Marcano-Velazquez et al., 2019). Metabolic engineering of another cellulolytic *Clostridium*, i.e., *C. cellulolyticum*, started earlier (namely in 2002) (Guedon, Desvaux, & Petitdemange, 2002) but less progress has been reported, so far. Work engineering metabolic pathways in *C. cellulovorans* started in 2015 (Yang, Xu, & Yang, 2015), and since then at least other seven studies have been published, demonstrating the great interest in this microbial model, especially for direct fermentation of plant biomass to *n*-butanol (Bao, Zhao, Li, Liu, & Yang, 2019; Wen, Ledesma-Amaro, Lin, Jiang, & Yangd, 2019; Wen, Ledesma-Amaro, Lu, Jin, & Yang, 2020).

The following sections will illustrate progress made in metabolic engineering of *C. thermocellum* for: (i) expanding the range of growth substrates and; (ii) improving the production of fuels and chemicals including ethanol, butanol, isobutanol, lactic acid and medium-chain esters. It is worth noting the significant interest in improving the ability of *C. thermocellum* to tolerate inhibitory compounds released from biomass by thermochemical pretreatment (e.g., furfural and hydroxymethylfurfural) (Kim, Groom,

Chung, Elkins, & Westpheling, 2017; Kim & Westpheling, 2018; Linville, Rodriguez, Brown, Mielenz, & Cox, 2014; Wilson et al., 2013). Nonetheless, continuous ball milling treatment during biomass fermentation, termed cotreatment, has recently been indicated as a promising alternative to thermochemical pretreatment for augmenting the capability of *C. thermocellum* to deconstruct plant biomass (Balch et al., 2017). The latter study reported that ball milling cotreatment enabled total carbohydrate solubilization of 88% for senescent switchgrass fermented by *C. thermocellum*. This new technology could therefore avoid toxicity issues resulting from thermochemical pretreatments in the future.

2. Overview of *C. thermocellum* central metabolism

C. thermocellum ferments cellulose, cellobiose (and other cellodextrins) and glucose (Lynd et al., 2002) through an atypical Embden-Meyerhof-Parnas pathway (Fig. 1) producing a mixture of products which include ethanol, H_2 and a number of organic acids (mainly acetate, formate, pyruvate and lactate) (Fig. 2)(Olson et al., 2017). *C. thermocellum* glycolysis shows several unique characteristics that affect its thermodynamic equilibrium (Fig. 1) (Jacobson et al., 2020). In particular, the conversion of fructose-6-phosphate (F6P) to fructose-1,6-bisphosphate (FBP), catalyzed by phosphofructokinase (Pfk), is coupled to phosphoryl group transfer from pyrophosphate (PP_i) instead of ATP. In addition, glucokinase (Glk; Glc +NTP→G6P) employs GTP rather than ATP as the high-energy phosphate donor, and phosphoglycerate kinase (Pgk; BPG+NDP→3PG +NTP) is capable of producing both ATP and GTP (Xiong et al., 2018; Zhou et al., 2013). Finally, *C. thermocellum* lacks a pyruvate kinase (Pyk) (Zhou et al., 2013) (Fig. 1). Instead, phosphoenolpyruvate (PEP) is converted to pyruvate, i.e., the final product of glycolysis, by two parallel pathways: (i) one, which can account for about 2/3 of the carbon flux, is catalyzed by pyruvate phosphate dikinase (Ppdk) and consists in the conversion of PEP to pyruvate with concomitant production of ATP; (ii) the other one has been called the "malate shunt" and consists of three reactions catalyzed by PEP carboxykinase (Pepck), producing oxaloacetic acid (OAA), NADH-dependent malate dehydrogenase (Mdh), which reduces OAA to malic acid (MA) and finally NADP-dependent malic enzyme (Me) which oxidatively decarboxylates MA to pyruvate (Deng et al., 2013; Olson et al., 2017; Zhou et al., 2013) (Fig. 1). The net effect of the malate shunt is that conversion of PEP to pyruvate is coupled to the production of GTP

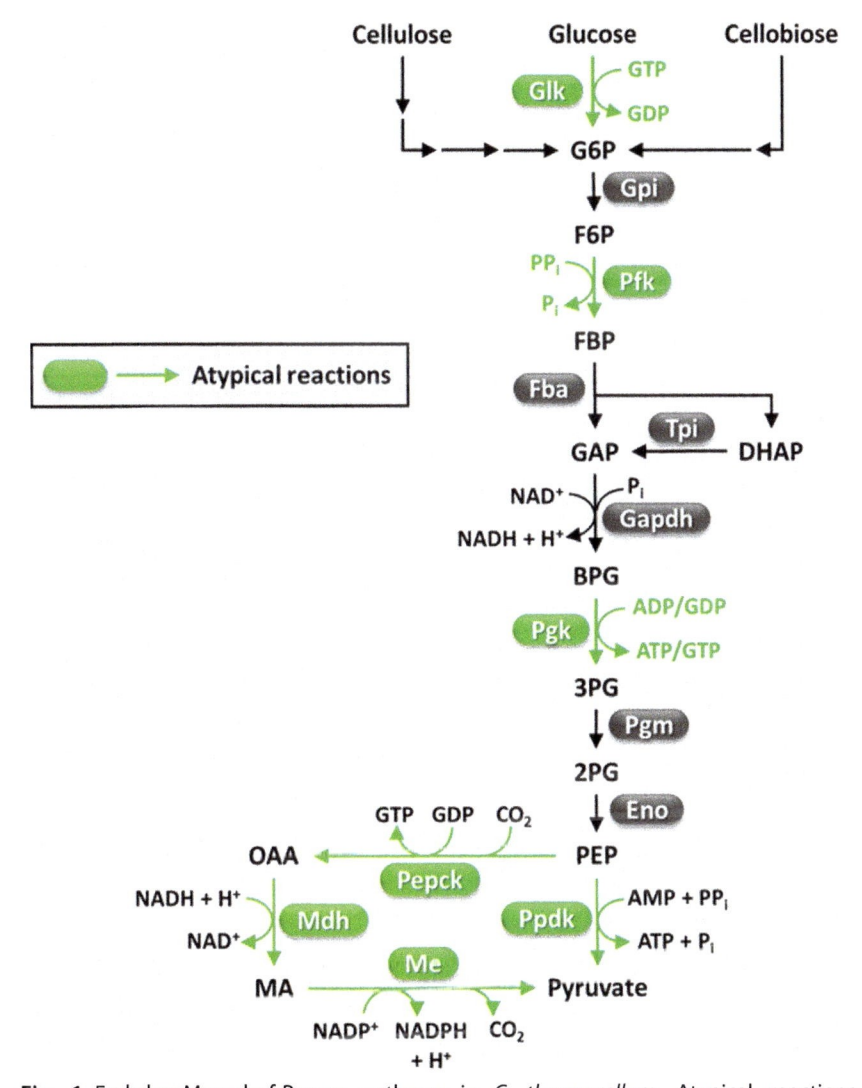

Fig. 1 Embden-Meyerhof-Parnas pathway in *C. thermocellum*. Atypical reactions are highlighted in green. Abbreviations: 2PG, 2-phosphoglyceric acid; 3PG, 3-phosphoglyceric acid; DHAP, dihydroxyacetone phosphate; Eno, enolase; F6P, fructose-6-phosphate; Fba, fructose 1,6-bisphosphate aldolase; FBP, fructose 1,6-bisphosphate; G6P, glucose-6-phosphate; GAP, glyceraldehyde 3-phosphate; Gapdh, glyceraldehyde 3-phosphate dehydrogenase; Glk, glucokinase; Gpi, glucose-6-phosphate isomerase; MA, malic acid; Mdh, malate dehydrogenase; Me, malic enzyme; OAA, oxalacetic acid; PEP, phosphoenolpyruvate; Pepck, phosphoenolpyruvate carboxykinase; Pfk, phosphofructokinase; Pgk, phosphoglycerate kinase; Pgm, phosphoglycerate mutase; PP_i, pyrophosphate; Ppdk, pyruvate phosphate dikinase; Tpi, triosephosphate isomerase.

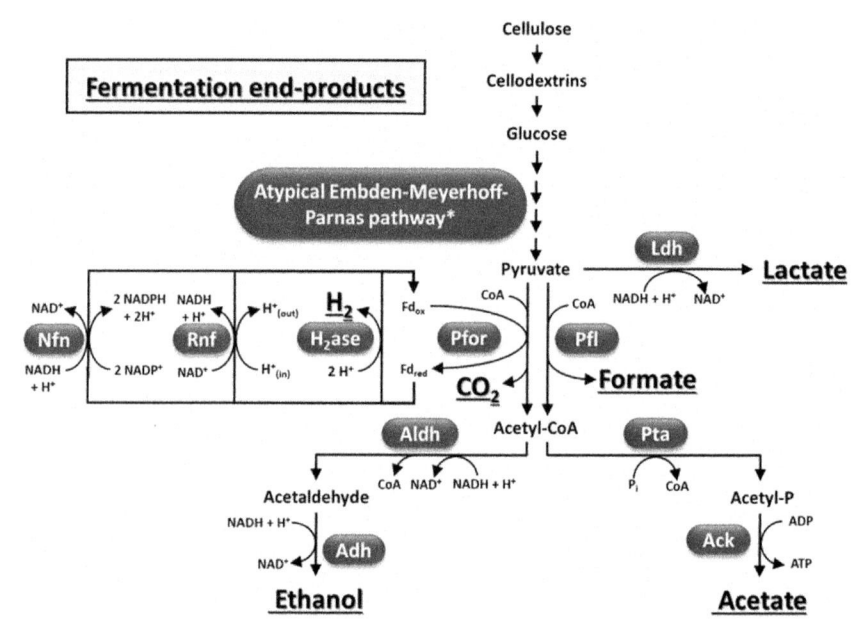

Fig. 2 Overview of *C. thermocellum* central metabolism. Fermentation end-products are underlined. *Please refer to text and Fig. 1 for detailed information on Embden-Meyerhof-Parnas pathway in *C. thermocellum*. Abbreviations: Acetyl-P, acetyl phosphate; Ack, acetate kinase; Adh, alcohol dehydrogenase; Aldh, aldehyde dehydrogenase; Fd, ferredoxin; H$_2$ase, hydrogenase; Ldh, lactate dehydrogenase; Nfn, NADH-dependent reduced ferredoxin: NADP$^+$ oxidoreductase; Pfl, pyruvate-formate lyase; Pfor, pyruvate: ferredoxin oxidoreductase; Pta, phosphotransacetylase; Rnf, ion-translocating reduced ferredoxin: NAD$^+$ oxidoreductase.

from GDP and NADH is exchanged for NADPH (via the differing cofactor specificities of the MDH and ME reactions). Pyruvate can be either reduced to lactate by lactate dehydrogenase (Ldh) or converted to acetyl-CoA by pyruvate ferredoxin oxidoreductase (Pfor) or pyruvate formate lyase (Pfl, which also produces formate) (Fig. 2). Finally, acetyl-CoA has two main metabolic fates: (i) conversion to acetate; (ii) reduction to ethanol (Deng et al., 2013). The first pathway includes acetyl-CoA conversion to acetyl-P by phosphotransacetylase (Pta), followed by acetyl-P conversion to acetate, with concomitant transfer of phosphate on ADP thus producing ATP, by acetate kinase (Ack). The ethanol biosynthetic pathway is a two-step reductive pathway catalyzed by the bifunctional alcohol/aldehyde dehydrogenase AdhE in which acetyl-CoA is reduced to acetaldehyde and then to ethanol by consumption of two NADH equivalents.

Apart from carbon metabolism, significant attention has also been focused on *C. thermocellum* electron metabolism because of its importance in the production of industrially relevant compounds such as ethanol and lactic acid (Lo et al., 2017; Tian, Lo, et al., 2016). The activity of the enzymes involved in the biosynthesis of these chemicals in *C. thermocellum* (i.e., alcohol dehydrogenase, Adh, aldehyde dehydrogenase, Aldh, and lactate dehydrogenase, Ldh) depends on nicotinamide cofactors. However, half of the electrons deriving from sugar catabolism are transferred to ferredoxin in the reaction catalyzed by Pfor. A number of reactions can be used to transfer electrons from reduced ferredoxin to nicotinamide cofactors (i.e., NAD^+ and $NADP^+$), and these are collectively known as ferredoxin:NAD oxidoreductase (FNOR) reactions. Some specialized versions of which include, ion-translocating reduced ferredoxin: NAD^+ oxidoreductase (RNF, Biegel, Schmidt, González, & Müller, 2011), and NADH-dependent reduced ferredoxin: $NADP^+$ oxidoreductase (NFN, Cui, Olson, & Lynd, 2019; Wang, Huang, Moll, & Thauer, 2010).

An additional target that might be worth investigating in the future is the global redox-responsive protein Rex (Sander et al., 2019; Zheng, Lanahan, Lynd, & Olson, 2018). Rex acts as a gene transcription repressor in response to low intracellular $NADH/NAD^+$ ratio (Ravcheev et al., 2012) and deletion of its gene determinant(s) led to accumulation of reduced fermentation end-products (mainly ethanol but also lactic acid) in anaerobic cellulolytic bacteria, such as *C. bescii* (Sander et al., 2019) and *Thermoanaerobacterium saccharolyticum* (Zheng et al., 2018), or in other clostridia, such as *C. pasteurianum* (Schwarz et al., 2017) or *C. acetobutylicum* (Wietzke & Bahl, 2012). The genomes of both *C. thermocellum* strain DSM 1313 and strain ATCC 27405 contain two genes annotated as encoding Rex (Clo1313_1799 and Clo1313_2471 and Cthe_0422 and Cthe_1798, respectively). The Clo1313_1799 coding sequence is thought to be the primary *rex* gene in *C. thermocellum* DSM 1313. Several attempts to delete it have failed (unpublished data). The Rex binding site consensus sequence for Clostridiales (5'-TTGTTAANNNNTTAACAA) reported by Ravcheev et al. (2012) has been generally used to predict Rex binding sites in the genome of clostridia (Hu et al., 2016; Schwarz et al., 2017). Previous studies on other clostridial models have nonetheless indicated that only limited confidence can be granted to these predictions and confirmation by experimental evidence is necessary (Hu et al., 2016).

Compared to other industrially relevant microorganisms such as *T. saccharolyticum*, *Escherichia coli* or *Saccharomyces cerevisiae* (in which glycolysis is much more exergonic), *C. thermocellum* glycolysis operates much closer to thermodynamic equilibrium (Jacobson et al., 2020). This is likely the result of evolutionary adaption to cellulolytic lifestyle, where substrate is abundant and energy is limiting. However, from an industrial application standpoint this is problematic, since this kind of metabolism is highly susceptible to product feedback inhibition (Jacobson et al., 2020). Several studies have attempted to identify the most limiting steps in *C. thermocellum* glycolysis and improve carbon flux through this pathway (Dash et al., 2019; Deng et al., 2013; Olson et al., 2017). Inactivation of the malate shunt and expression of a heterologous Pyk increases *C. thermocellum* growth rate by about 30% (Olson et al., 2017). Due to carbon fixation in the PEPCK reaction (Fig. 1), flux through the malate shunt is sensitive to CO_2 levels. Low levels of dissolved CO_2 may present a thermodynamic barrier to flux (Dash et al., 2019). This may explain the observation that growth of *C. thermocellum* is stimulated by the presence of 10%–20% CO_2 in the headspace. Although under certain conditions, the PPDK reaction can support 2/3 of PEP to pyruvate flux, it does not seem to be capable of supporting *all* of the flux, since the malate shunt cannot be disrupted, except in the presence of a heterologous *pyk* gene (Deng et al., 2014). Fba and Gapdh have been flagged as other probable enzymes subject to inhibition by product feedback (Dash et al., 2019). Although their inhibition by accumulation of sugar phosphates (Fba) or NADH (Gapdh) has been studied with particular reference to high ethanol concentration, it is possible that accumulation of other fermentation end-products of industrial interest (e.g., lactic acid) may exhibit similar effects. Based on this evidence, improving glycolytic flux in *C. thermocellum* seems a feasible and desirable target for development of higher performing strains for industrial applications.

3. Improving fermentation of hemicellulose-derived sugars by *C. thermocellum*

Apart from cellulose (which generally represents 40%–50% of lignocellulose), plant biomass comprises other different polymers: hemicellulose, pectin and lignin, whose abundance varies widely depending on the plant species, but generally ranges around 25%–20% for hemicellulose plus pectin and 15%–20% for lignin (Gray et al., 2006). Extensive bioconversion of all these components is essential for 2nd generation biorefining processes to be cost competitive.

One of the main limits of *C. thermocellum* as candidate for use in a pure culture CBP system is its inability to grow on the main fraction of hemicellulose (i.e., xylan, although it can grow on mixed-linkage β-glucans such as barley β glucan and laminarin) and pectin (Aburaya et al., 2015; Demain et al., 2005; Fuchs, Zverlov, Velikodvorskaya, Lottspeich, & Schwarz, 2003). Ideally, CBP should target all the carbohydrate fractions of plant biomass. It is worth noting that *C. thermocellum* biosynthesizes several xylanases, even when grown on cellobiose (Raman et al., 2009), whose activity was demonstrated in vitro (Morag, Bayer, & Lamed, 1990). It has therefore been speculated that lack of growth on xylan is related to the inability of *C. thermocellum* to metabolize xylose and/or xylose oligomers (Xiong, Reyes, Michener, Maness, & Chou, 2018). Genes for xylose uptake and metabolism have been recently identified, but their activity causes intracellular accumulation of xylitol (Verbeke et al., 2017). Furthermore, xylose, several xylooligosaccharides and other pentoses exert inhibitory effects of *C. thermocellum* growth. Engineering pentose catabolism in *C. thermocellum* would therefore also benefit cellulose breakdown by alleviating inhibition from C5 sugars (Verbeke et al., 2017).

The first effort to engineer *C. thermocellum* for xylose consumption involved expressing the xylose isomerase (*xi*) and xylulose kinase (*xk*) genes from *T. saccharolyticum* in a mini-operon at the *ldh* locus. The *pta* gene was then disrupted to eliminate acetate production. Initially, the strain grew poorly in xylose. Subsequent serial transfer improved the growth rate. This strain was able to produce about 2 g/L ethanol from 10 g/L xylose, however its only public description is a patent application, and thus is not well known (Argyros, Barrett, Caiazza, & Hogsett, 2011). Independently, the *xi* and *xk* genes from *Thermoanaerobacter ethanolicus* were integrated into the *C. thermocellum* chromosome under the control of a strong promoter (the *C. thermocellum* glyceraldehyde 3-phosphate dehydrogenase promoter, P_{Gapdh}) (Xiong, Reyes, et al., 2018). This engineered strain can co-utilize a pentose (xylose) with hexoses including glucose, cellobiose, and crystalline cellulose without carbon catabolite repression. Ethanol was produced at a titer of 14 mM (0.6 g/L) from a mixture of 2.5 g/L xylose and 2.5 g/L Avicel cellulose. Furthermore, the engineered strain could ferment xylooligomers (degree of polymerization comprised between 2 and 7) at least to a certain extent by exploiting the native *C. thermocellum* endoxylanase activity (Xiong, Reyes, et al., 2018). However, these tests indicated that β-xylosidase activity (i.e., conversion of xylo-trimers and dimers to xylose) is another rate-limiting step in this strain, unless some xylose is already present in the medium. It has been hypothesized that induction of the native β-xylosidase and/or xylose transporters could be induced only by xylose.

Although this progress is promising, the ability of *C. thermocellum* to use the hemicellulose fractions of lignocellulosic biomass currently seems underdeveloped for industrial applications. *C. cellulolyticum* can ferment a larger spectrum of substrates, since it can grow on both cellulose and xylan (Saxena, Fierobe, Gaudin, Guerlesquin, & Belaich, 1995). *C. cellulovorans* can ferment all the major plant polysaccharides, including cellulose, xylan, galactomannan and pectin (Aburaya et al., 2015). Additional research is therefore needed to improve utilization of hemicellulose and pectin by *C. thermocellum*. However, it is also worth remembering that co-culture of *C. thermocellum* with thermophilic hemicellulolytic microorganisms (e.g., *T. saccharolyticum*, *Thermoanaerobacterium thermosaccharolyticum* and *Herbinix* spp.) has proved to be a valuable alternative approach to efficiently fermenting both the cellulose and hemicellulose components of plant biomass (Beri, York, Lynd, Peña, & Herring, 2020; Froese, Schellenberg, & Sparling, 2019; He, Hemme, Jiang, He, & Zhou, 2011; Jiang et al., 2013). Although designing and maintaining a stable microbial consortium may be challenging at the industrial scale (Johns, Blazejewski, Gomes, & Wang, 2016), this remains an attractive option for improving the efficiency of lignocellulosic fermentation by *C. thermocellum*.

4. Improving production of high-value chemicals in *C. thermocellum* by metabolic engineering

4.1 Ethanol

Humans have produced ethanol by fermentation for thousands of years (Legras, Merdinoglu, Cornuet, & Karst, 2007). Ethanol is the most widely produced biofuel today, with over 16 billion gallons produced in the United States alone in 2019 ("Ethanol industry outlook. Powered with renewed energy", 2019). Its high oxygen content (relative to gasoline) allows it to replace toxic alternatives such as methyl *tert*-butyl ether (MTBE) and tetraethyllead (TEL) for increasing octane rating and preventing engine knock. It is usually blended with gasoline for use in spark-ignited engines (10% in the United States and 27% in Brazil). Since 2003, automobile engine technology has been available that allows use of blends ranging from 20% to 100% ethanol. Since the 1980s, diesel cycle engines (i.e. compression ignited), have been produced that can use a blend of 95% ethanol called ED95 (https://www.sekab.com/en/products-services/product/ed95/). Ethanol can also be catalytically upgraded to longer-chain fuel molecules, such as jet fuel and gasoline, often at a cost similar to that of distillation (Hannon et al., 2020). 3.1 EJ/year of ethanol is currently produced from sugar cane (30%)

and grains (50%), but projected future global demand of bioethanol for transport will require a significant contribution by cellulosic bioethanol (Lynd et al., 2017). Life cycle analysis has indicated that reduction of greenhouse gas emission of 19%–48% could be obtained by replacing gasoline with corn ethanol, while a reduction up to 115% could be attained by using cellulosic ethanol (Wang, Han, Dunn, Cai, & Elgowainy, 2015). Currently, bioethanol is mainly obtained by fermentation of mono- or disaccharides using *S. cerevisiae* or *Zymomonas mobilis*. However, these microorganisms cannot directly ferment lignocellulosic biomass and require expensive processes for biomass pretreatment and hydrolysis (Himmel et al., 2007). This has inspired work to engineer *C. thermocellum* for improved ethanol production (Fig. 3).

4.1.1 Pathways competing for carbon and electron flux

C. thermocellum naturally produces ethanol but at low yield, typically 12%–34% of the maximum theoretical (that is 2 mol/mol of glucose equivalent) (Olson, Sparling, & Lynd, 2015). However, this is lower than the titer of 40 g/L and a yield of 90% of theoretical maximum that are generally considered as necessary for commercial applications (Dien, Cotta, & Jeffries, 2003).

Initial attempts to engineer *C. thermocellum* for increased ethanol production focused on disruption of pathways competing for carbon and electron flux, namely acetate, lactate, formate, and hydrogen (Fig. 3). The first of these targets was acetate (Tripathi et al., 2010). Wild type *C. thermocellum* produces acetate and ethanol in a ratio that varies from 3:1 to 1:1 under normal conditions (Argyros, Tripathi, et al., 2011; Holwerda et al., 2020; Papanek, Biswas, Rydzak, & Guss, 2015) although it can be higher in the presence of active hydrogen removal when co-cultured with a methanogen (Weimer & Zeikus, 1977). Acetate is produced from acetyl-CoA via the combined action of phosphotransacetylase (*pta*) and acetate kinase (*ack*). Deletion of one (Argyros, Tripathi, et al., 2011; Tripathi et al., 2010) or both genes (Papanek et al., 2015) has been shown to eliminate more than 95% of acetate production. Initially, cell growth is inhibited, and carbon flux is diverted to pyruvate. After serial transfer or adaptation for faster growth in a chemostat, flux to pyruvate decreases. In most cases, this results in an increase in ethanol (Argyros, Tripathi, et al., 2011), although in some cases, the flux was largely redirected to amino acid production (van der Veen et al., 2013). The shift in flux from pyruvate to ethanol appears to be due to a D494G mutation in the bifunctional aldehyde dehydrogenase (ALDH) and alcohol dehydrogenase (ADH), *adhE*, which increases NADPH-linked

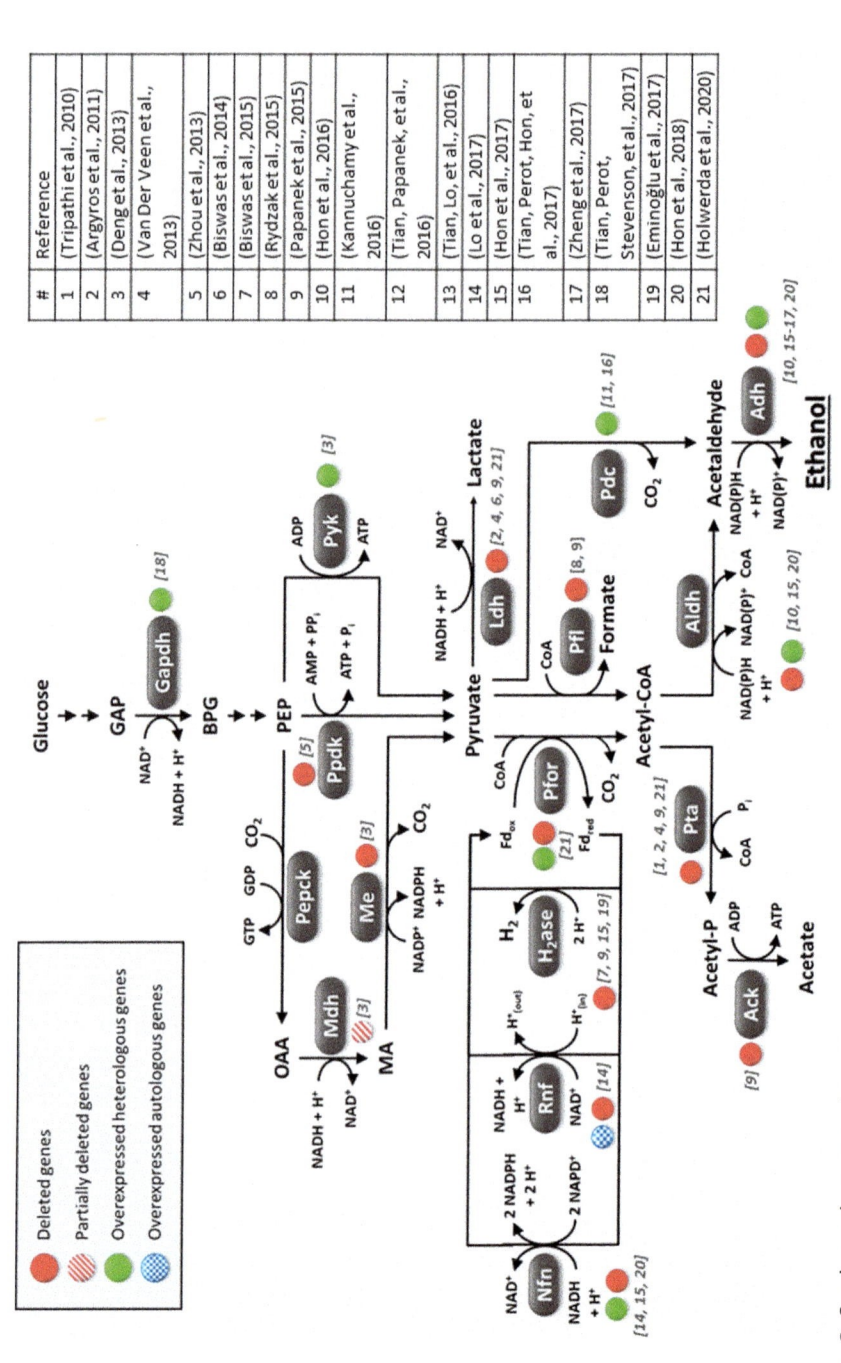

Fig. 3 See legend on opposite page.

ADH activity (Holwerda et al., 2020; Zheng et al., 2015). Taken together, these results strongly suggest that ethanol production is controlled by electron availability (in the form of NADH or NADPH) rather than carbon availability (i.e., acetyl-CoA).

Lactate production has also been disrupted to increase ethanol production in *C. thermocellum* (Fig. 3). In the wild type organism, lactate production is part of the overflow metabolism response. When normal pathways for producing acetate and ethanol are impaired, several metabolites in central metabolism increase. An increase in levels of FBP is the signal that activates Ldh which produces lactate from pyruvate (Lo, Zheng, Hon, Olson, & Lynd, 2015; Özkan, Yilmaz, Lynd, & Özcengiz, 2004). Under typical fermentation conditions, *C. thermocellum* produces low levels of lactate, however under conditions of high substrate loading, low pH, or in strains where acetate (Argyros, Tripathi, et al., 2011) or ethanol production (Lo, Zheng, Hon, et al., 2015) has been disrupted, lactate levels can be much higher. Combined elimination of both lactate and acetate pathways resulted in strains capable of producing high titers of ethanol (25–30 g/L), although at relatively modest yields (50%–60% of the theoretical maximum) (Holwerda et al., 2020). In these strains, overflow metabolites included isobutanol (2–5 g/L), valine (5–7 g/L), alanine (1–3 g/L), pyroglutamate (1–4 g/L), and fumarate (1–4 g/L).

Hydrogen and formate compete with ethanol as sinks for electrons, and thus eliminating one or both is another strategy to increase ethanol yield (Fig. 3). The pyruvate formate lyase (PFL) reaction converts pyruvate to acetyl-CoA with the concomitant generation of formate. Deletion of the

Fig. 3 Overview of the main gene modification strategies used to improve ethanol production in *C. thermocellum*. Red dots, genes that were deleted; red dashed dots, genes that were partially deleted; green dots, overexpressed heterologous genes; blue checkered dots, overexpressed autologous genes. Numbers in square brackets refer to studies in which gene modifications were performed (listed in the table). Abbreviations: Acetyl-P, acetyl phosphate; Ack, acetate kinase; Adh, alcohol dehydrogenase; Aldh, aldehyde dehydrogenase; BPG, 1,3-bisphospho glycerate; Fd, ferredoxin; GAP, glyceraldehyde 3-phosphate; Gapdh, glyceraldehyde 3-phosphate dehydrogenase; H$_2$ase, hydrogenase; Ldh, lactate dehydrogenase; MA, malic acid; Mdh, malate dehydrogenase; Me, malic enzyme; Nfn, NADH-dependent reduced ferredoxin: NADP$^+$ oxidoreductase; OAA, oxaloacetic acid; Pdc, pyruvate decarboxylase; PEP, phosphoenolpyruvate; Pepck, phosphoenolpyruvate carboxy kinase; Pfl, pyruvate-formate lyase; Pfor, pyruvate:ferredoxin oxidoreductase; PP$_i$, pyrophosphate; Ppdk, pyruvate phosphate dikinase; Pta, phosphotransacetylase; Pyk, pyruvate kinase; Rnf, ion-translocating reduced ferredoxin: NAD$^+$ oxidoreductase.

pfl gene eliminated formate production but had very little effect on ethanol production. It did, however, have a substantial effect on growth, decreasing the growth rate of *C. thermocellum* by 80%. This growth defect could be restored by supplementation with 2 mM formate, suggesting that *pfl* may play a role in C1 metabolism in this organism (Rydzak, Lynd, & Guss, 2015). Deletion of the *hydG* hydrogenase maturation protein and *ech* energy-conserving hydrogenase eliminated all hydrogen production in *C. thermocellum* (Biswas, Zheng, Olson, Lynd, & Guss, 2015). Ethanol yield increased to 64% of the theoretical maximum. During construction of the *hydG* deletion, the strain acquired the D494G mutation in the *adhE* gene. Untangling the relative contributions of the D494G mutation vs. *hydG* deletion is difficult because all strains with the *hydG* deletion also have the D494G mutation. The D494G mutation by itself (strain LL1161) increases ethanol yield from 26% to 44% of the theoretical maximum, and under similar fermentation conditions, produces more ethanol than the *hydG* deletion strain (LL350) (Zheng et al., 2015), thus the D494G mutation appears to be the dominant effect. A separate study looked at deletion of the *hfsB* subunit of the *hfs* hydrogenase. This led to a moderate increase in ethanol yield (increase from 37% to 55% of the theoretical maximum), although this is thought to be due to a regulatory effect of this deletion on ADH activity, rather than a direct effect on hydrogen production (Eminoğlu et al., 2017).

Combining deletions of *hydG*, *pfl*, *ldh* and *pta-ack* resulted in a strain that produced ethanol at high yield (78% of theoretical maximum) at low substrate loadings (1 g/L cellobiose), but the yield rapidly decreased as substrate loading increased (Papanek et al., 2015). Adapting the strain for faster growth on low (5 g/L) substrate increased its ability to consume substrate but resulted in glucose accumulation. Subsequent adaptive evolution at high (50 g/L) substrate reduced glucose accumulation and increased ethanol production. The resulting strain produced ethanol at a yield of 75% of theoretical maximum and maximum titer of 27 g/L (although at this titer, the yield was slightly lower) (Tian, Papanek, et al., 2016). Analysis of protein expression data showed an increase in a putative operon (Clo1313_1831-33) which has a gene annotated as a phosphofructokinase B. In addition, the metabolic proteins AdhE and Gapdh showed a twofold increase compared to the starting strain (AG553), although it is not clear what may have caused this increase. Although adaptive evolution increased the growth rate of the final strain (LL1210), its growth rate was still only about half that of the wild type strain (0.22 vs 0.53 h^{-1}).

4.1.2 Electron flux from ferredoxin

To produce ethanol at yields greater than 50% of the theoretical maximum, electrons from ferredoxin need to be transferred to ethanol. This transformation is accomplished by a class of enzymes known as ferredoxin nicotinamide oxidoreductases (FNORs). Transfer of electrons from ferredoxin to NAD^+ or $NADP^+$ is highly exergonic, so this reaction is often coupled to an endergonic reaction for energy conservation (Buckel & Thauer, 2013). In *C. thermocellum*, the transfer of electrons from ferredoxin to NAD^+ is coupled to proton pumping across the membrane by the Rnf enzyme. The transfer of electrons from ferredoxin to $NADP^+$ is coupled to transhydrogenation (NADH converted to NADPH) by the NfnAB enzyme (Lo et al., 2017). Disruption of RNF activity (by deletion of the entire *rnf* operon or just the D and G subunits) generally causes a decrease in ethanol production. By contrast, overexpression is more variable, causing ethanol production to increase, decrease or remain unchanged, depending on the genetic background. One possible explanation is that if the RNF reaction operates close to equilibrium, increased enzyme dosage may not have a large effect on net flux (Noor et al., 2014). Another possible explanation is that in mutants where the ADH reaction is not strictly NADH-linked, the ability of the RNF reaction to generate NADH is less important. Deletion of *nfnAB* has a negligible role on ethanol production in all strains tested so far. Like RNF, the NFN reaction may operate close to equilibrium, and in some cases may function primarily to eliminate excess NADPH (Olson et al., 2017). The inability to generate a double mutant of *ppdk* and *nfnAB*, despite successfully creating both single mutants, supports this hypothesis (unpublished data).

4.1.3 Disrupting the malate shunt

A unique feature of *C. thermocellum* glycolysis is that it does not have a canonical pyruvate kinase (Pyk) enzyme for conversion of PEP to pyruvate. Instead, it uses either pyruvate phosphate dikinase (Ppdk) or a set of three enzymes known as the malate shunt, that convert PEP to oxaloacetate, malate, and then pyruvate (Olson et al., 2017). The PYK reaction has a large negative Gibbs free energy, and is thought to be essentially irreversible under physiological conditions, which allows it to provide a strong driving force for reactions downstream of pyruvate. By contrast, both the PPDK and malate shunt reactions are thought to be more reversible (Dash et al., 2019). Thus, one approach for increasing ethanol production in *C. thermocellum* is to replace these pathways with a heterologous *pyk* gene. Heterologous expression of the *pyk* gene from *T. saccharolyticum* resulted in only a small

(~10%) increase in ethanol yield (Deng et al., 2013). Subsequent disruption of the malate shunt further increased ethanol yield to about 47% of theoretical (Deng et al., 2014), with a maximum titer of 15 g/L (unpublished data). Disruption of the malate shunt also led to the appearance of an I480K mutation in the heterologous Pyk protein (Olson et al., 2017). The effect of this mutation is unknown. Deleting *ppdk* had no effect on ethanol production, although it has a large effect on the distribution of intracellular fluxes.

4.1.4 The linkage between nitrogen metabolism and ethanol metabolism

Apart from canonical fermentation products, *C. thermocellum* also secretes relevant (4%–10%) amino acid amounts, especially valine and alanine, which are sometimes increased by certain gene modifications (e.g., 17% in evolved Δ*ldh* Δ*pta* strains) (van der Veen et al., 2013). In order to reduce amino acid synthesis by *C. thermocellum*, ammonium assimilation was diminished by deleting *glnA* that encodes a Type I glutamine synthetase (Rydzak et al., 2017). Levels of secreted valine and total amino acids were reduced by 53% and 44% in the Δ*glnA* strain, respectively, and this resulted in enhanced ethanol yields by 53%. However, since carbon flux to secreted amino acids is only about 10% of carbon flux to ethanol in the wild type, the observed increase in ethanol cannot be accounted for simply by decreased carbon flux to amino acids.

The conversion of PEP to pyruvate can occur via two pathways (PPDK and the malate shunt, as described above). The control of these pathways appears to be regulated by intracellular concentrations of PP_i and ammonium (NH_4^+). Malic enzyme is activated by ammonium (Lamed & Zeikus, 1981) and inhibited by PP_i (Taillefer, Rydzak, Levin, Oresnik, & Sparling, 2015). PPDK activity is activated by ammonium (Olson et al., 2017), and uses PP_i as a cofactor. Low concentrations of ammonium should inhibit both pathways, limiting flux to pyruvate, which provides a plausible explanation for the connection between nitrogen and carbon metabolism. Since *glnA* is responsible for ammonium assimilation (i.e., combining intracellular ammonium and glutamate to produce glutamine), its deletion could potentially result in higher intracellular ammonium levels, although this remains to be determined experimentally. This mechanism for increasing ethanol production (i.e., deletion of *glnA* increases intracellular ammonium levels, which activates pathways for PEP to pyruvate conversion) could be further tested by introducing the *glnA* deletion into a strain where PEP is converted to pyruvate exclusively by PYK (i.e., strain LL1251, Olson et al., 2017). If the ammonium activation hypothesis is true, we would expect a *glnA* deletion to have no effect on the ethanol production of strain LL1251.

4.1.5 Heterologous expression of genes for ethanol production

T. saccharolyticum is a thermophilic anaerobic bacterium that has been engineered to produce ethanol at high yield (80%–90% of theoretical) and high titer (~70 g/L) from hexose and pentose sugars. It does not have the ability to ferment cellulose, however, so there has been interest in transferring the ethanol producing ability of this bacterium to *C. thermocellum*. Initially, genes from the pyruvate to ethanol pathway were targeted for transfer. This includes the NADPH-linked ADH, *adhA*, the bifunctional ALDH and ADH, *adhE*, the NADPH-linked FNOR, *nfnAB*, ferredoxin, and pyruvate ferredoxin oxidoreductase (*pforA*), which were identified over the course of a series of gene deletion studies in *T. saccharolyticum* (Lo, Zheng, Olson, et al., 2015; Lo, Zheng, Hon, et al., 2015; Zheng et al., 2017; Zhou et al., 2015) (Fig. 3). Of these, expressing *adhA* by itself increased ethanol yield from 20% to 35% of the theoretical maximum. In the presence of *adhA*, expressing *nfnAB* (from *T. saccharolyticum*) further increased ethanol yield to 45% of theoretical. Expressing *adhE* (from *T. saccharolyticum*) had no effect on ethanol yield (Hon et al., 2017). Interestingly, the benefits of expressing the *T. saccharolyticum* ethanol production pathway genes were not additive with previous genetic modifications, including the *hydG* deletion, *ldh* and *pta* deletions, and malate shunt disruption (in fact, the addition of the pathway genes actually reduced ethanol yield). Chromosomal expression of the *T. saccharolyticum* ethanol production pathway worked better than plasmid-based expression, increasing ethanol yield from 31% to 74% of theoretical. Expression of *T. saccharolyticum pforA* in *C. thermocellum* did not have an effect at low substrate loading, but increased final ethanol titer from 220 to 350 mM at higher substrate loading (52 g/L cellobiose) (Hon et al., 2018).

In addition to the pyruvate to ethanol pathway genes, two genes from *T. saccharolyticum* further upstream from pyruvate have been expressed in *C. thermocellum*. One (*pyk*) was mentioned in a previous section. The other, glyceraldehyde-3-phosphate dehydrogenase (*gapdh*) was added after observing a metabolic bottleneck at the GAPDH reaction when wild type *C. thermocellum* was grown in the presence of increasing concentrations of added ethanol (Tian, Perot, Stevenson, et al., 2017). The Gapdh protein from *T. saccharolyticum* appears to be less inhibited by high NADH/NAD$^+$ ratios, compared to the native Gapdh protein of *C. thermocellum*.

The main effect of this genetic modification is to decrease lag time in the presence of added ethanol (Tian, Perot, Stevenson, et al., 2017). In some cases, we have observed a slight (~5%) improvement in maximum ethanol titer (unpublished data).

In general, attempts to express genes from *T. saccharolyticum* in *C. thermocellum* have worked well. One notable exception is the NADH-linked FNOR gene, *tsac_1705*. This gene was only successfully expressed in strains of *C. thermocellum* where *rnf* had been deleted, suggesting that expression of both *tsac_1705* and *rnf* may result in a futile cycle that disrupts the membrane proton gradient (Tian, Lo, et al., 2016).

Frequently, understanding the effect of heterologous gene expression is made difficult by the presence of native enzymes with the same function. To better understand the function of the heterologous gene, we have in some cases deleted the corresponding native genes. The *adhE* gene from *C. thermocellum* has been deleted in both the wild type strain and a strain expressing *T. saccharolyticum* pyruvate to ethanol pathway genes. In the wild type strain, the *T. saccharolyticum adhE* gene was able to fully complement deletion of the native *adhE* gene (Hon et al., 2016). However, it does not appear to be able to support high yield and titer ethanol production. In the presence of the rest of the *T. saccharolyticum* pyruvate to ethanol pathway, deletion of the native *adhE* reduced the yield from 74% to 33% of theoretical (Hon et al., 2017). In the presence of the *T. saccharolyticum pforA* gene, all five native *pfor* gene clusters in *C. thermocellum* could be deleted without any decrease in ethanol yield or titer (in fact, the yield slightly increased because deletion of *pfor4* eliminated isobutanol production) (Hon et al., 2018).

Although most of the genes expressed in *C. thermocellum* have been from *T. saccharolyticum*, one exception is pyruvate decarboxylase (*pdc*). The PDC reaction mediates the conversion of pyruvate to acetaldehyde, skipping acetyl-CoA, which is an intermediate in the native pathway (Fig. 3). The PDC pathway is used by organisms such as *S. cerevisiae* and *Z. mobilis* for commercial ethanol production. An interesting property of most Pdc enzymes is that although the purified enzymes are fairly resistant to high temperature conditions (Raj, Talarico, Ingram, & Maupin-Furlow, 2002), the upper temperature of this pathway seems to be around 45–48° C (Ishchuk et al., 2008; Kata, Semkiv, Ruchala, Dmytruk, & Sibirny, 2016; Van Zyl, Taylor, Eley, Tuffin, & Cowan, 2013).

Recently there have been two attempts to express a *pdc* gene in *C. thermocellum*. In one example, the gene from *Z. mobilis* was expressed, causing ethanol yield to increase from 26% to 53% of *pdc* theoretical (although only low substrate loadings were tested) (Kannuchamy, Mukund, & Saleena, 2016). In another example, the *pdc* gene from *Acetobacter pasteurianus* was shown to increase the theoretical yield from 40% to 75% of theoretical, but only in the presence of a heterologous *adhA* from *T. saccharolyticum* (in the absence of *adhA*, ethanol yield increased, but by a much smaller amount). The increased ethanol yield is puzzling because PDC activity was 100-fold lower than expected (Tian, Perot, Hon, et al., 2017).

4.1.6 Other methods of identifying metabolic limitations in C. thermocellum

In addition to experiments designed to directly improve ethanol yield and/or titer, several experiments have been performed to identify factors limiting ethanol production in *C. thermocellum*. In one experiment, cells were grown in the presence of gradually increasing amounts of added ethanol. The resulting increase in metabolites upstream of the GAPDH reaction (glucose-6-phosphate, fructose-1,6-bisphosphate, dihydroxyacetone phosphate) and depletion of metabolites downstream of the GAPDH reaction (3-phosphoglycerate and PEP) suggested a metabolic bottleneck at that reaction (Tian, Perot, Stevenson, et al., 2017).

Another experiment looked at the ability of *C. thermocellum* enzymes in the pyruvate to ethanol pathway to support high yield and titer ethanol production in *T. saccharolyticum*. The *adhE*, *nfnAB*, and *ferredoxin* genes from *C. thermocellum* were sufficient both individually and in combination. The *C. thermocellum pfor1* gene was able to support high yield ethanol production, but only at low titers. At the highest substrate loading (160 g/L), the *C. thermocellum pfor1* could only support 300 mM ethanol titer, compared to 1100 mM with the *T. saccharolyticum pforA* (Cui et al., 2019).

A final approach for understanding metabolic limitations in *C. thermocellum* is through the use of a cell-free system. In this system, cells were lysed, and the lysate was used to show conversion of cellobiose to ethanol. Cellobiose was rapidly converted to hexose phosphates (glucose-6-phosphate and fructose-6-phosphate). FBP appeared much more slowly, indicating a potential metabolic bottleneck at the phosphofructokinase (PFK) reaction, but not excluding other downstream bottlenecks. An interesting feature of this system was its ability to function at 37 °C. This allowed testing the addition of commercially available mesophilic enzymes to this system. Of all the enzymes tested, Adh from yeast showed the largest improvement in ethanol production. Surprisingly, the addition of Pdc (also from yeast) did not improve ethanol production, and in fact decreased it. One possible interpretation is that *C. thermocellum* enzymes are particularly sensitive to inhibition by acetaldehyde (Cui et al., 2020).

4.1.7 Adaptive evolution experiments

Adaptive laboratory evolution (ALE) is frequently used to modify microbial phenotypes. A prerequisite for ALE is the ability to link improvements to the phenotype of interest with improvements in growth rate. Product formation can be linked to growth by gene deletions. This strategy was pursued in the adaptive evolution of strains with deletions of *ldh* and *pta* (Holwerda et al., 2020). In three of four evolved lineages, ethanol production was improved. In the fourth strain (LL375), ethanol production did not improve. Instead, the strain found a way to increase its growth rate improved despite high levels of pyruvate and amino acid secretion. Eliminating these pathways through additional gene deletions would be necessary for further improvements with this strain. This strategy was also pursued in the adaptive evolution of strain LL1210 (deletions of *hydG*, *pfl*, *ldh*, and *pta-ack*, described above), and resulted in improved yield and titer, first by increasing the total amount of substrate that could be consumed, and then by diverting flux from glucose to ethanol (Tian, Papanek, et al., 2016). The final growth rate of this strain is still only about half that of the wild type strain, suggesting that further improvements in ethanol production may be possible.

Since both wild type and engineered strains of *C. thermocellum* seem to stop producing ethanol when titers reach 25–30 g/L (either due to ethanol production or external addition), we suspect that ethanol tolerance may limit titer. Ethanol tolerance has been improved in wild type *C. thermocellum* to 50–80 g/L (Brown et al., 2011; Shao et al., 2011; Williams, Combs, Lynn, & Strobel, 2007). In several of these cases, mutations were found in the *adhE* gene, and in one case, the *adhE* mutations alone were sufficient to recapitulate the ethanol tolerance phenotype. However, ethanol tolerance sometimes comes at a cost of decreased ethanol production. Reintroduction of the P704L and H734R mutations in AdhE (collectively known as AdhE*) diverted carbon flux from ethanol to lactate production, although subsequent deletion of the *ldh* gene was able to restore ethanol production to wild type levels (Biswas, Prabhu, Lynd, & Guss, 2014). In another example, a complete deletion of *adhE* (which eliminated ethanol production) increased ethanol tolerance from 15 to 30 g/L (Tian, Cervenka, Low, Olson, & Lynd, 2019).

4.1.8 AdhE is a hotspot for mutations in C. thermocellum

In strains of *C. thermocellum* engineered for increased ethanol production or adapted for increased ethanol tolerance, the *adhE* gene is a hotspot for mutations (Fig. 4). With the exception of the heterologous *pyk* from *T. saccharolyticum* (described above), we have not observed point mutations in any of the genes that participate in the pathway from cellobiose to ethanol except *adhE*. However in *adhE*, we have identified *seven* distinct mutations, and have performed some characterization on three of them. The D494G mutation has been shown to increase NADPH-linked ADH activity, which results in increased ethanol titer (Hon et al., 2016). The pair of mutations, P704L and H734R, reduce product inhibition from ethanol and NAD^+ for the ADH reaction. When both mutations are present in *C. thermocellum*, they confer increased ethanol tolerance, but reduce ethanol titer (Biswas et al., 2014). Kinetic properties and ethanol tolerance phenotypes of the remaining four AdhE mutations identified in *C. thermocellum* have not been measured.

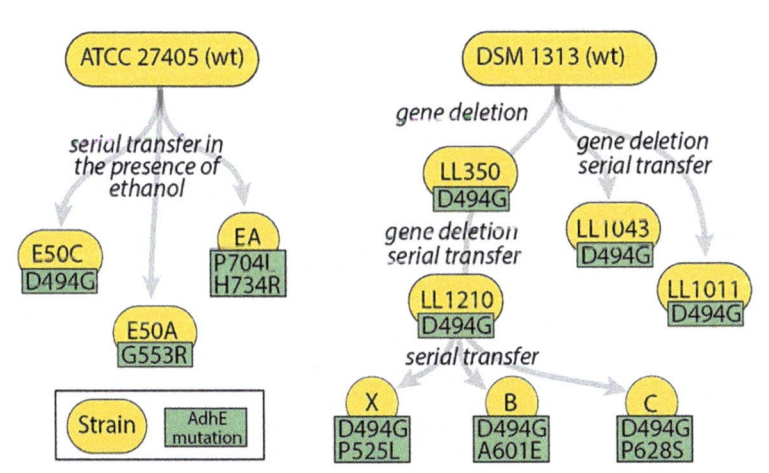

Fig. 4 Lineages of strains of *C. thermocellum* where AdhE mutations have been observed. Strain ATCC 27405 is the type strain. Strain DSM 1313 is another strain which is more genetically tractable than ATCC 27405 and is thus the background for all existing genetically modified strains of *C. thermocellum*. The *adhE* sequence in both ATCC 27405 and DSM 1313 is identical. The D494G mutation is unique in that it has evolved independently four times (three times in DSM 1313 lineages and once in an ATCC 27405 lineage).

4.1.9 Role of excess NADPH production in existing metabolic engineering efforts

Eliminating the malate shunt, expressing *adhA*, and the D494G mutation in *adhE* all have a similar effect with respect to increased ethanol yield. Furthermore, when these mutations have been combined, there was no additive effect on ethanol production (Hon et al., 2017), suggesting epistatic effects. One possible explanation is that under normal conditions, flux through the malate shunt results in excess production of NADPH, beyond what is needed for biosynthesis. In the wild type strain, this excess NADPH can be eliminated either by reverse flux through NFN (with a corresponding increase in hydrogen production), or via amino acid production (Olson et al., 2017). The introduction of either *adhA* (from *T. saccharolyticum*) or the D494G mutation in the native *adhE* gene results in an increase in NADPH-linked ADH activity, thus allowing excess NADPH to be eliminated by increased ethanol production. When the malate shunt is disrupted, no excess NADPH accumulates, and more NADH is available for ethanol production. From a theoretical perspective, using an NADPH-linked ADH may be preferable to eliminating the transhydrogenation activity of the malate shunt because in this case, increases in ethanol titer will not affect the $NADH/NAD^+$ ratio (instead affecting the $NADPH/NADP^+$ ratio) and thus will not inhibit the GAPDH reaction. This reaction operates near thermodynamic equilibrium in many organisms (Noor et al., 2014) which makes it particularly susceptible to product inhibition.

4.1.10 Remaining metabolic limitations in C. thermocellum

Given the substantial effort that has gone into engineering *C. thermocellum* for improved ethanol production over the past decade, it is good to review our knowledge of options for future improvements. Broadly speaking, there have been two main approaches to improving ethanol production, deleting competing pathways (resulting in strain LL1210) (Tian, Papanek, et al., 2016) and expressing heterologous genes from *T. saccharolyticum* (resulting in strain LL1570) (Hon et al., 2018). Interestingly, both approaches have resulted in strains with similar yield (∼75%–80% of theoretical) and titer (∼25 g/L). Attempting to combine the mutations by introducing the *T. saccharolyticum* pathway into strain LL1210 was unsuccessful (Hon et al., 2017). Attempts to delete pathways for competing end products in strains with the *T. saccharolyticum* pyruvate to ethanol pathways have met with modest success (*hydG*, *ech* and *ldh* have been deleted, *pta-ack* was tried with no success, *pfl* has not been attempted, Hon et al., 2017, and unpublished data), however in this strain, further increases in ethanol titer appear to be limited by either accumulation of glucose or lack of substrate consumption.

4.2 n-Butanol and isobutanol

n-Butanol (1-butanol) has a longer carbon backbone than ethanol (4 carbons vs 2), which gives it fuel properties more similar to that of gasoline, such as high combustion energy, low volatility and low corrosivity. Pure *n*-butanol can be fed to spark ignited engines without any modification, while ethanol must be blended with gasoline (Campos-Fernández, Arnal, Gómez, & Dorado, 2012). Isobutanol has energy density similar to n-butanol and higher octane number, which is an advantage for blending into gasoline (Chen & Liao, 2016). Furthermore, isobutanol can be dehydrated to form isobutene, which can then be oligomerized to C8 then C12 alkenes to be used as jet fuel (Lin et al., 2015). Despite these advantages, *n*-butanol exhibits higher toxicity compared to ethanol (Ingram, 1976) which limits fermentation titers, and separation (by distillation) typically requires two distillation columns instead of one (Vane, 2008), potentially increasing capital costs.

The most well-known pathway for *n*-butanol production is the native one found in mesophilic solvent-producing Clostridia (Fig. 5), such as *C. acetobutylicum* and *C. beijerinckii*, where it leads to *n*-butanol titers of 15–20 g/L (Chen & Liao, 2016; Nicolaou, Gaida, & Papoutsakis, 2010; Tomas, Welker, & Papoutsakis, 2003). This pathway has been engineered in a number of heterologous hosts such as *E. coli*, *S. cerevisiae*, *Lactobacillus brevis* and *Pseudomonas putida* (Atsumi, Hanai, & Liao, 2008; Berezina et al., 2010; Nielsen et al., 2009; Steen et al., 2008), resulting in limited production of n-butanol (final titer lower than 1 n g/L). A much larger success was obtained by replacing the native butyryl-CoA dehydrogenase/electron transfer protein (Bcd/EtfAB) with the ferredoxin-independent trans-enoyl-CoA reductase (Ter) for reduction of crotonyl-CoA to butyryl-CoA at least in *E. coli* (Shen et al., 2011). This synthetic pathway enabled *E. coli* to produce n-butanol titers (up to 30 n g/L) similar to those accumulated by the native n-butanol producers but it did not have the same effectiveness in other hosts such as in n*S. cerevisiae* (Lian, Si, Nair, & Zhao, 2014). Apart from heterologous *ter* expression, the successful example of n-butanol production in n*E. coli* was obtained also through: (i) deletion of *pta* (which eliminated acetyl-CoA consumption toward acetate) and; (ii) increasing the cellular NADH pool by disrupting all the alternative fermentative pathways

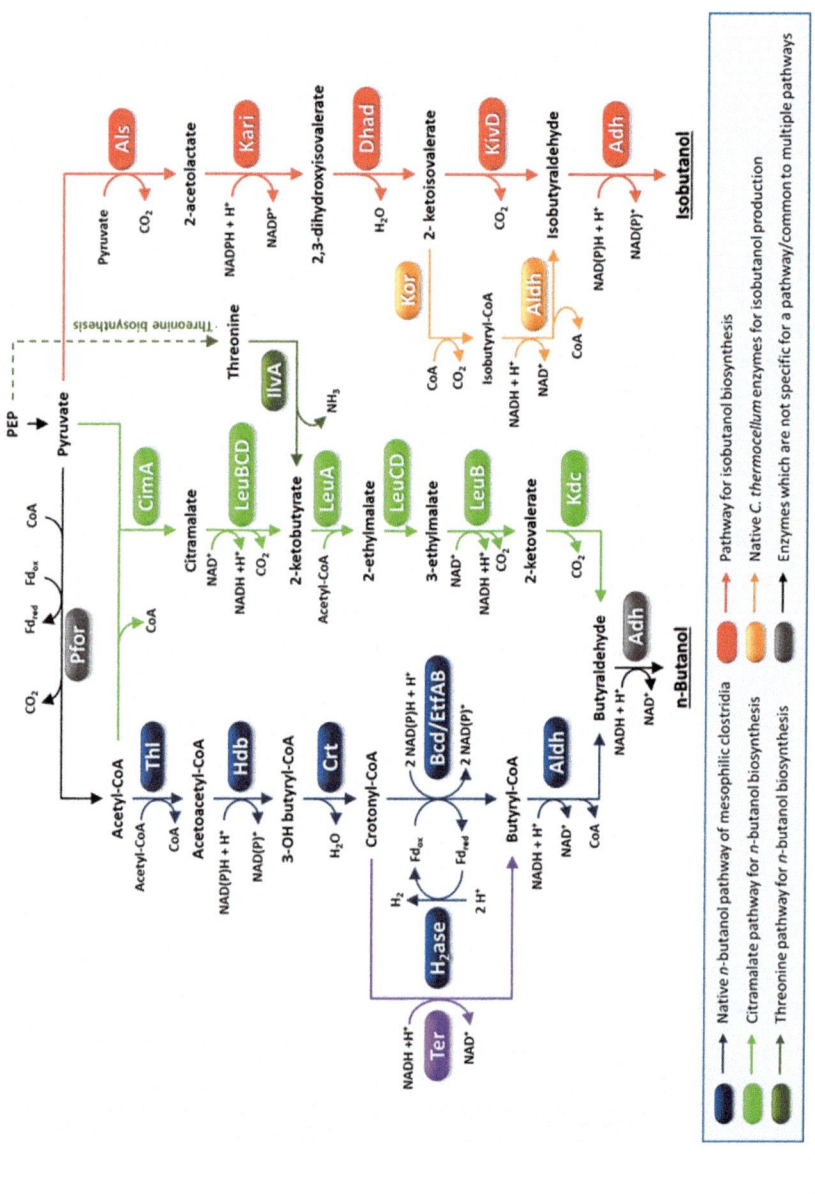

Fig. 5 See legend on opposite page.

competing for NADH and expressing a formate dehydrogenase (that oxidizes formate with production of NADH) (Shen et al., 2011). These additional modifications created accumulation of acetyl-CoA and NADH which acted as metabolic driving forces that made n-butanol synthesis thermodynamically more favorable. This is important because condensation of acetyl-CoA to acetoacetyl-CoA by thiolase (Thl) is particularly difficult from a thermodynamic perspective (nFig. 5). Alternatively, 2-keto acid pathway(s) can be engineered to produce n-butanol through biosynthesis of 2-ketobutyrate, followed by its decarboxylation and reduction (nFig. 5) (Shi et al., 2016).

Engineering of valine biosynthetic pathway, involving other 2-keto acids (i.e., pyruvate, 2-keto isovalerate) was used to produce isobutanol (Atsumi et al., 2008). Although decarboxylation of pyruvate drives this intermediate toward ethanol biosynthesis, longer chain (C > 3) 2-keto acids are generally used as precursors in amino acid biosynthesis. However, native pathways for isobutanol production were recently identified (Gu et al., 2017; Holwerda et al., 2014). In *Klebsiella pneumoniae*, only disruption of competing pathways (i.e., α-acetolactate decarboxylase) led to actual accumulation of isobutanol in this strain (Gu et al., 2017). Interestingly, wild-type *C. thermocellum* is able to produce detectable amounts (1.6 g/L) of isobutanol (Holwerda et al., 2014). A recent study discovered that *C. thermocellum* Pfor4 is necessary for isobutanol production (Hon et al., 2018). In order to further increase isobutanol titer, the challenge is to engineer flux through the 2-keto acid pathways toward longer chain alcohols. Production of isobutanol from renewable sources has been engineered in

---

Fig. 5 Pathways for *n*-butanol and isobutanol biosynthesis. The native *n*-butanol pathway of mesophilic solventogenic clostridia is indicated in blue. For more favorable thermodynamics, when this pathway has been engineering in heterologous microorganisms, the Bcd/EftAB complex has often been replaced with Ter (purple). Citramalate and threonine pathways for *n*-butanol biosynthesis are indicated in light green and dark green, respectively. Red indicates pathway for isobutanol biosynthesis, while orange was used for native *C. thermocellum* enzymes involved in isobutanol production. Abbreviations: Adh, alcohol dehydrogenase; Aldh, aldehyde dehydrogenase; Als, α-acetolactate synthase; Bcd/EftAB, butyryl-CoA dehydrogenase/electron transfer protein; CimA, citramalate synthase; Crt, crotonase; Dhad, dihydroxy acid dehydratase; Fd, ferredoxin; H$_2$ase, hydrogenase; Hbd, 3-hydroxybutyryl-CoA dehydrogenase; Kari, keto acid reductoisomerase; Kdc, 2-ketoacid decarboxylase; KivD, 2-ketoisovalerate decarboxylase; Kor, ketoisovalerate ferredoxin-dependent reductase; IlvA, threonine dehydratase; LeuA, 2-isopropylmalate synthase; LeuB, 3-isopropylmalate dehydrogenase; LeuCD, isopropylmalate isomerase; PEP, phosphoenolpyruvate; sPfor, pyruvate ferredoxin/flavodoxin oxidoreductase; Ter, trans-enoyl-CoA reductase; Thl, thiolase.

a number of microbial models (Atsumi et al., 2008; Atsumi, Higashide, & Liao, 2009; Higashide, Li, Yang, & Liao, 2011; Li et al., 2012; Lin et al., 2014; Smith, Cho, & Liao, 2010). This evidence suggests that engineering isobutanol production is compatible with most organisms and may be easier compared to *n*-butanol also thanks to a larger number of pathway enzymes available from a variety of sources (Chen & Liao, 2016). The thermodynamics of the isobutanol pathway are more favorable than the *n*-butanol pathway due to a strong negative Gibbs free energy for all of the pathway steps.

Few thermophilic *n*-butanol pathways have been reported so far (Tian et al., 2019). This dramatically limits the number of genes available for butanol engineering in *C. thermocellum*. Nonetheless, 12 different *n*-butanol pathway permutations were recently assembled in *C. thermocellum* (Tian, Conway, et al., 2019). This led to development of butanol producing *C. thermocellum* strains, but the titer obtained was limited to less than 1 g/L (Tian, Conway, et al., 2019). Consistent with similar studies on other microbial models, the efficiency of *n*-butanol pathways engineered in *C. thermocellum* seems limited at multiple levels that include gene expression, enzyme stability, co-factor availability, and reaction thermodynamics (Tian, Conway, et al., 2019). Through optimization of pathway gene combination, choice of transcriptional promoter, choice of gene locus for integration, disruption of interfering pathways (i.e., lactic acid and isobutanol production), protein engineering (i.e., Thl, 3-hydroxybutyryl-CoA dehydrogenase, Hbd, and Ter), and process engineering (i.e., pH regulation, ethanol supplementation) 357 mg/L *n*-butanol was produced from 50 g/L crystalline cellulose in 5 days (Tian, Conway, et al., 2019). Issues with heterologous expression of *n*-butanol pathway enzymes are intrinsically connected to the complexity of this pathway. However, this can also be seen as a set of opportunities for its future improvement. It is worth noting that although wild-type *C. thermocellum* can tolerate only 5 g/L *n*-butanol, adaptive evolution improved tolerance up to 15 g/L (Tian, Cervenka, et al., 2019). Currently, tolerance does not limit *n*-butanol production in *C. thermocellum*, although further improvement of its n-butanol tolerance may be required in the future.

Improvement of isobutanol production in *C. thermocellum* by metabolic engineering was first reported in 2015 (Lin et al., 2015). This study took advantage of a set of thermophilic isobutanol pathway enzymes identified in a previous study on another thermophilic bacterium, namely *Geobacillus thermoglucosidasius* (Lin et al., 2014). This study highlighted toxicity of enzymes involved in isobutanol biosynthesis, and in particular of genes encoding acetolactate synthase (Als). A large number of gene constructs, combining different promoters for each gene, was necessary to find the right balance between insufficient expression of the pathway genes leading to low isobutanol titer, and excessive expression leading to cell toxicity. In optimized culture conditions, the most efficient engineered *C. thermocellum* strain produced 5.4 g/L of isobutanol from cellulose within 75 h (the wild type strain could produce only 1.5 g/L in the same conditions), corresponding to 41% of theoretical yield (Lin et al., 2015). This titer closely matches the isobutanol tolerance of the wild type *C. thermocellum* (Tian, Cervenka, et al., 2019), suggesting that isobutanol production may, in fact, be limited by tolerance. The adaptive evolution procedure used to improve *C. thermocellum* tolerance to n-butanol also increased tolerance to isobutanol to the same extent (i.e., 15 n g/L) (Tian, Cervenka, et al., 2019). Interestingly, deletion of *adhE* was even more effective, leading to a strain that can grow with an initial concentration of isobutanol of 17.5 g/L (Tian, Cervenka, et al., 2019). These achievements will possibly provide help for future further enhancement of isobutanol production in *C. thermocellum*. However, it is currently not clear if the last steps of isobutanol production pathway (i.e., reduction of isobutyryl-CoA to isobutanol) rely on AdhE, so that deleting *adhE* gene would also be detrimental to isobutanol production.

Cellulosic *n*-butanol production was first reported by using engineered mesophilic cellulosic clostridia (Gaida, Liedtke, Jentges, Engels, & Jennewein, 2016; Yang et al., 2015). Furthermore, a recently isolated *T. thermosaccharolyticum* has shown native ability to ferment microcrystalline cellulose to *n*-butanol (Li, Zhang, Yang, & He, 2018). Research on *C. cellulovorans* is currently leading this field also owing to the fact that this strain is naturally equipped with a butyrate biosynthetic pathway (Usai et al., 2020). Recently, about 4 g/L of *n*-butanol was obtained by batch fermentation of cellulose with *C. cellulovorans* expressing *adhE2* gene encoding a bifunctional aldehyde/alcohol dehydrogenase from *C. acetobutylicum* (Bao et al., 2019). Moreover, further modification of *C. cellulovorans* metabolism (including relief of carbon catabolite repression on pentose metabolism, overexpression of heterologous Ter and butyryl-CoA-acetate CoA transferase) led to a strain able to ferment alkali-extracted corn cobs to 4.96 g/L *n*-butanol in 97 h (Wen, Ledesma-Amaro, Lu, Jin, & Yang, 2020). Titers obtained by using engineered *C. cellulovorans* currently are even higher that those obtain with native cellulosic n-butanol producer *nT. thermosaccharolyticum* (1.93 g/L through fermentation of microcrystalline cellulose) (Li et al., 2018). *C. cellulovorans* is currently positioned far ahead of *C. thermocellum* in terms of final titer, yield and even productivity of *n*-butanol production (Table 1) although the best example of non-native

Table 1 Summary of the C. thermocellum strains engineered so far that most efficiently produce ethanol, n-butanol, isobutanol, lactic acid or isobutyl acetate through direct fermentation of crystalline cellulose, compared with the most performant recombinant cellulolytic organisms found in the literature.

Product	Strain	Strain identifier	Growth substrate	Y (mol/mol hexose equivalent) (% of the maximum theoretical)	Final titer (g/L)	Productivity (g/L·h⁻¹)	Reference
Ethanol	C. thermocellum DSM 1313 Δhpt^a $\Delta hydG$ Δldh Δpfl Δpta-ack $adhE^{D494G}$, evolved	LL1210	95 g/L Avicel	1.53 (75%)	27	≈0.2	Tian, Papanek, et al. (2016)
	C. thermocellum DSM 1313 Δhpt^a. $\Delta pfor1$–5, $\Delta Clo1313\_1483$, Tsac_adhA, $adhE^{G544D}$, Tsac_fd, Tsac_nfnAB, Tsac_pforA	LL1570	100 g/L Avicel	1.08 (54%)	25	0.70	Hon et al. (2018)
	C. thermocellum DSM 1313 Δhpt^a, Δldh, Δpta, $adhE^{D494G}$, evolved	LL1011	120 g/L Avicel	≈1.15 (≈58%)	29.9	≈0.15	Holwerda et al. (2020)
	Saccharomyces cerevisiae with codon-optimized cellulase genes encoding T. emersonii CbhI, T. reesei EglII and A. aculeatus BglI	SK13-34	10 g/L Avicel	1.5 (75%)	3.8	0.04	Song et al. (2018)
n-Butanol	C. thermocellum DSM 1313 Δhpt^a, $\Delta Clo1313\_0478^b$, Δldh, with gene Tt_thl, Tt_hbd, Tt_crt, St_ter, Ts_bad and Ts_bdh integrated in the genome, $\Delta Clo1313\_1353$–1356^c, Tt_thlM2^d, Tt_hbdM^e, St_terM^f	LL1668	50 g/L Avicel + 4 g/L ethanol	0.02 (2%)	0.357	0.004	Tian, Conway, et al. (2019)
	C. cellulovorans 743B Ca_adhE2	adhE2	40 g/L cellulose	0.53 (53%)	4.0	≈0.013	Bao et al. (2019)

Isobutanol	*C. thermocellum* DSM 1313 Δ*hpt*[a] *Ll_kivD*, *Ct_ilvBN*, *Ct_ilvCD*	CT24	80 g/L Avicel	0.39 (39%)	5.4	0.072	Lin et al. (2015)
	C. cellulolyticum Bs_*alsS*, *Ec ilvCD*, *Ll_kivD*, *Ec_yqhD*	—	10 g/L crystalline cellulose	n.d.	0.660	0.003	Higashide et al. (2011)
Lactic acid	*C. thermocellum* DSM 1313 Δ*hpt*[2] Δ*adhE ldh*[S161R]	LL1111	5 g/L cellobiose[g]	0.78 (39%)	2.02	n.d.	Lo, Zheng, Hon, et al. (2015)
	Caldicellulosiruptor besci P*xi*-*ldh*	MACB1066	5 g/L xylose[g]	≈ 0.35 (≈ 18%)	≈ 0.5	≈ 0.02	Williams-Rhaesa et al. (2018)
Isobutyl acetate	*C. thermocellum* DSM 1313 Δ*hpt*[a] mutant *Sa_cat* ΔClo1313_0613 ΔClo1313_0693[h,i]	HSCT2105	20 g/L Avicel	2.8×10^{-4} (0.04%)[j]	0.0031	4.3×10^{-5}	Seo, Nicely, and Trinh (2020)

[a]Disruption of the *hpt* gene in *C. thermocellum* DSM 1313 was performed to allow for 8AZH counter-selection, and it does not affect the fermentation phenotype (Argyros, Tripathi, et al., 2011).

[b]Gene *Clo1313_0478* encodes a native restriction system of *C. thermocellum* (Hon et al., 2017).

[c]Genes *Clo1313_1353–1356* are involved in isobutanol biosynthesis (Hon et al., 2018).

[d]*T_t_thlM2* is a mutant version of the *Thermoanaerobacter thermosaccharolyticum* thiolase with the following changes R133G, H156N, P222F and N223V.

[e]*T_t_hbdM*, is a variant of the *T. thermosaccharolyticum* 3-hydroxybutyryl-CoA dehydrogenase with the following changes D31A, I32R, and P36I.

[f]*St_terM* is a mutant version of the *Spirochaeta thermophila* trans-enoyl-CoA reductase with the substitution of E75A.

[g]Growth of these strains on crystalline cellulose was not determined.

[h]Genes Clo1313_0613 and Clo1313_0693 encode two carbohydrate esterases of *C. thermocellum* (Seo et al., 2020).

[i]As far as we know this is the only example of direct fermentation of cellulose to isobutyl acetate.

[j]The maximum theoretical yield is 0.67 mol/mol hexose equivalent.

The symbol "≈" was used for approximate values that were calculated from data in the corresponding studies. n.d., not determined. Abbreviations: *ak*, acetate kinase; *adhA*, alcohol dehydrogenase; *adhE*, bifunctional aldehyde/alcohol dehydrogenase; *alsS*, α-acetolactate synthase; *bad*, butyraldehyde dehydrogenase; *bdh*: alcohol dehydrogenase; Bgl, β-glucosidase; Bs, *Bacillus subtilis*; Ca, *Clostridium acetobutylicum*; *cat*, chloramphenicol acetyl-transferase; Cbh, cellobiohydrolase; *crt*, crotonase; Ct, *Clostridium thermocellum*; Ec, *Escherichia coli*; Eg, endoglucanase; *fld*, ferredoxin; *hbd*, 3-hydroxybutyryl-CoA dehydrogenase; *hydG*, maturase of the three [Fe—Fe] hydrogenases of *C. thermocellum*; *ilvBN*, acetohydroxy acid synthase; ilvC, keto acid reductoisomerase; *kivD*, dihydroxy acid dehydratase; *ldh*, lactate dehydrogenase; Ll, *Lactococcus lactis*; *tfnAB*, NADH-dependent reduced ferredoxin: NADP+ oxidoreductase; *pfl*, pyruvate formate lyase; *pforA*, pyruvate ferredoxin/flavodoxin oxidoreductase; *pta*, phosphotransacetylase; P*xi*, xylose-inducible promoter; Sa, *Staphylococcus aureus*; St, *Spirochaeta thermophila*; *ter*, trans-enoyl-CoA reductase; *thl*, thiolase; Ts, *Thermoanaerobacter sp.* X514; *Tsac*, *Thermoanaerobacterium saccharolyticum*; Tt, *Thermoanaerobacter thermosaccharolyticum*; *yqhD*, alcohol dehydrogenase.

n-butanol pathway is that engineered in *E. coli* (n-butanol titer up to 30 g/L) (Shen et al., 2011). Apart from the choice of enzymes, which is nonetheless restricted by the availability of thermophilic ones, other factors that could significantly increase the efficiency of n-butanol pathway are increasing the intracellular levels of acetyl-CoA and the n)HNAD(P)H as previously reported in *E. coli* (Shen et al., 2011). In *E. coli*, an increase of the intracellular acetyl-CoA concentration was obtained by inactivating acetate biosynthesis (Shen et al., 2011). A number of options (e.g., elimination of H_2 and/or ethanol, improving FNOR activity) could be used to improve the intracellular concentration of HNAD(P)H in *C. thermocellum* by taking advantage of new understanding developed during attempts to improve ethanol production in this strain.

With respect to isobutanol, strains of *C. thermocellum* have demonstrated the highest titer production of this chemical from cellulose. As far as we know, isobutanol production has been engineered only in another cellulolytic strain, i.e., *C. cellulolyticum* (Higashide et al., 2011). However, this organism has lower growth and cellulose consumption rate than *C. thermocellum* (Lin et al., 2015) which results in lower isobutanol productivity and titer (Higashide et al., 2011).

4.3 Lactic acid

Lactic acid is one of the most requested chemicals worldwide (Alves de Oliveira, Komesu, Vaz Rossell, & Maciel Filho, 2018). Traditional lactic acid uses are in the food, cosmetic and pharmaceutical industries, but the application that is contributing the most to the current global market expansion of lactic acid is as building block for the synthesis of biodegradable plastic polymers (e.g., polylactide, PLA, and its co-polymers) (Abdel-Rahman, Tashiro, & Sonomoto, 2013; Alves de Oliveira et al., 2018). Lactic acid is currently produced mainly through fermentation of expensive food crops such as corn (Alves de Oliveira et al., 2018; Biddy, Scarlata, & Kinchin, 2016). However, economic and ethical concerns have promoted research on utilization of non-food feedstocks, especially lignocellulose. So far, research aimed at development of microorganisms for direct fermentation of lignocellulose to lactic acid have been mainly addressed at engineering cellulolytic characteristics (e.g., by expression of heterologous cellulases) in native lactic acid producers, such as lactic acid bacteria (Tarraran & Mazzoli, 2018). However, these strategies have had limited success owing to the intricacy of the native cellulose systems and to issues hampering their expression in heterologous hosts (Mazzoli, 2020). Looking for alternative strategies such as improvement of lactic acid production in cellulolytic microorganisms is therefore recommended and some evidence suggests that this approach is promising (Mazzoli, 2020).

Lactic acid yield in wild type *C. thermocellum* is very low (i.e., 0.01 mol/mol glucose equivalent), but single gene modification has been shown to dramatically improve it (Lo, Zheng, Hon, et al., 2015). The *C. thermocellum* strain that shows the highest lactic acid yield obtained so far is LL1111, which was engineered by disruption of the *adhE* gene encoding its main bifunctional alcohol/aldehyde dehydrogenase (Lo, Zheng, Hon, et al., 2015). Additionally, LL1111 features a spontaneous mutation of its *ldh* gene resulting in a Ldh whose activity is independent from allosteric activation by fructose 1,6 bisphosphate (FBP) (Lo, Zheng, Hon, et al., 2015). Lactic acid production yield in this strain is only about 40% of the theoretical maximum (i.e., 2 mol/mol glucose equivalent) while typical efficiencies of lactic acid producing microorganisms (e.g., lactic acid bacteria) which are used in industrial processes generally attain almost 100% conversion yield (with respect to available carbohydrates) (Mazzoli, Bosco, Mizrahi, Bayer, & Pessione, 2014). Apart from lactic acid, strain LL1111 produces significant amounts of other end-products, mainly acetate (i.e., 1 mol per mole of cellobiose, which corresponds to about 17% of the carbon flux) and H_2 (i.e., 2.7 mol per mole of cellobiose) (Lo, Zheng, Hon, et al., 2015). Further improvement of lactic acid production in *C. thermocellum* by engineering of its metabolic pathways is therefore necessary. Previous studies have suggested that lactic acid production in this strain is likely regulated at multiple levels, namely by: (i) competition with other pathways serving as alternative electron sinks (e.g., H_2 production) or diverting carbon intermediates elsewhere (e.g., acetate production) (Biswas et al., 2015; Lo, Zheng, Hon, et al., 2015); (ii) allosteric regulation of Ldh activity (Biswas et al., 2015); iii) modulation of the expression of the *ldh* gene (Ravcheev et al., 2012) (Fig. 6). However, these hypotheses need to be experimentally confirmed. Although the K_m for pyruvate of *C. thermocellum* Ldh was determined more than a decade ago (Özkan et al., 2004), no information on the affinity of the other enzymes that directly compete for pyruvate in *C. thermocellum* metabolism, i.e., Pfor and Pfl, is currently available. Among the five Pfors encoded by the *C. thermocellum* genome, some evidence indicates that *pfor1* (Clo1313_0020-0023) and *pfor4* (Clo1313_1353-1356) are those playing the main role in this strain (Hon et al., 2018; Rydzak et al., 2012; Xiong et al., 2016). *C. thermocellum* Pfl is encoded by *pflB* gene (Clo1313_1717) but its K_m for pyruvate has not been measured.

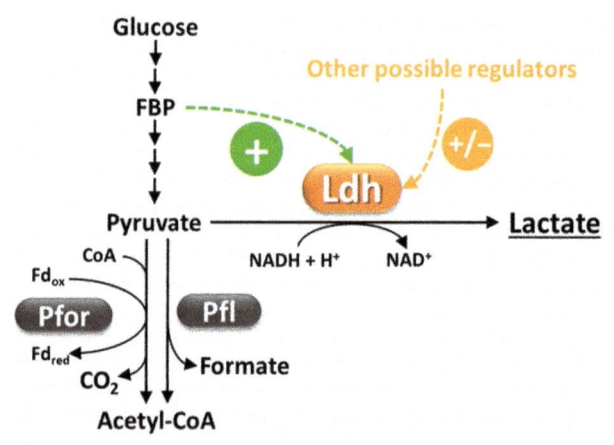

Fig. 6 LA production in *C. thermocellum* is regulated at multiple levels including regulation of expression of Ldh (orange), allosteric regulation of Ldh activity and competition for substrates and co-factors. FBP is the allosteric activator (+, green dashed arrow) of Ldh. Other factors [e.g., ATP, GTP, PP$_i$, NAD(P)$^+$, NAD(P)H and the global redox-responsive transcription factor Rex] may also affect Ldh activity and need to be tested (+/−, yellow dashed arrow). Furthermore, the K$_m$ of enzymes (Pfor, Pfl) that catalyze reactions that directly compete for pyruvate with Ldh is not known yet. Abbreviations: FBP, fructose 1,6 bisphosphate; Fd, ferredoxin; Ldh, lactate dehydrogenase; Pfl, pyruvate-formate lyase; Pfor, pyruvate ferredoxin/flavodoxin oxidoreductase.

With respect to allosteric modulation, it has been hypothesized that, apart from FBP, the *C. thermocellum* Ldh may also be activated by ATP and inhibited by PP$_i$ as was observed in *Caldicellulosiruptor saccharolyticus* (Willquist & van Niel, 2010). Nicotinamide cofactors [NAD(P)$^+$, NAD(P)H] are other typical regulators of Ldh (Bryant, 1991; Willquist & van Niel, 2010). As mentioned previously, *C. thermocellum* glycolysis shows several unique features such as utilization of guanine nucleotides instead of adenine nucleotides by Glk and Pgk (Jacobson et al., 2020). For this reason, testing possible allosteric effects of additional compounds such as GTP/GDP on Ldh activity is recommended. Additionally, in many microorganisms, *ldh* expression is under transcriptional regulation by the global redox-responsive transcription factor Rex (Ravcheev et al., 2012). Recently, we have shown that overexpression of Ldh can significantly improve lactic acid yield in *C. thermocellum*, highlighting the importance of this third level of regulation of lactic acid production in this strain (unpublished data). A systematic characterization of *C. thermocellum* Ldh regulation seems important to refine metabolic engineering strategies aimed at improving lactic acid production in this strain.

A further issue that hampers high lactic acid production in *C. thermocellum* is that this organism, similar to other anaerobic cellulolytic bacteria, is very sensitive to acids (Lynd et al., 2002; Whitham et al., 2018). *C. thermocellum* growth is severely limited at pH values around 6.0 (Whitham et al., 2018). Although some strong native lactic acid producers, such as lactic acid bacteria, can tolerate acidic pH as low as 3.2, growth inhibition by acidic pH and organic acid accumulation is a frequent cause of lactic acid productivity decrease also in these strains (Abdel-Rahman & Sonomoto, 2016; Alves de Oliveira et al., 2018). Hence, developing *C. thermocellum* strains with improved acidic pH/lactate tolerance is essential for reducing the cost of fermentative production of lactic acid from cellulose. Some strategies for improving acidic pH tolerance of *C. thermocellum* have been recently proposed such as: (i) improving the expression of F_1F_0-ATPase, (owing to its function in ATP-dependent pumping of protons out of the cell); (ii) upregulating proton-pumping PP_i-ase; (iii) improving the expression of protein chaperones and heat-shock proteins such as GrpE, Hsp20, and Hsp33; (iv) expressing a heterologous glutamate decarboxylase (involved in neutralizing pH acidity through proton-consuming decarboxylation of glutamate to γ–aminobutyrate); (v) inactivation of glutamine synthase (Whitham et al., 2018). However, the few examples aimed at improving acid tolerance in native cellulolytic microorganisms only include recent studies on *C. cellulovorans* (Wen et al., 2020) and *Fibrobacter succinogenes* (Wu et al., 2017). A combination of random chemical mutagenesis and evolutionary engineering has been used to increase acid tolerance in *F. succinogenes*. The improvement of acid tolerance in this strain was moderate since the pH limit was lowered from 6.10 to 5.65 (Wu et al., 2017). Adaptive evolution together with rational metabolic engineering allowed for a slightly larger improvement of acid tolerance in *C. cellulovorans* (Wen, Ledesma-Amaro, Lu, Jiang, et al., 2020). In the wild-type *C. cellulovorans*, growth at pH=6.5 causes more than 60% inhibition of biomass production, and almost no growth occurs at pH=6 (Usai et al., 2020). After adaptive evolution to more acidic pH, two cell wall lyase genes (Clocel_0798 and Clocel_2169, suspected to be involved in cell autolysis) were inactivated, and *augA* encoding agmatine deiminase (catalyzing hydrolytic removal of amino group from agmatine and releasing N-carbamoylputrescine and ammonia) from *C. beijerinckii* NCIMB 8052 was overexpressed (Wen, Ledesma-Amaro, Lu, Jiang, et al., 2020). The generated strain could tolerate a pH of 5.5. These studies represent promising paradigms possibly inspiring similar strategies on *C. thermocellum*.

4.4 Medium-chain esters

Recently, increasing attention has been given to microorganisms as potential platforms for eco-friendly production of medium-chain (C6-C10) esters (Kruis et al., 2019; Layton & Trinh, 2014; Park, Shaffer, & Bennett, 2009; Rodriguez, Tashiro, & Atsumi, 2014). These compounds show promise as drop-in biofuels and have additional industrial applications such as food additives, lubricants and solvents (Kruis et al., 2019; Lange et al., 2010). Traditional synthesis of esters is performed by the Fisher esterification process which requires high energy input and a strong acid (Liu, Lotero, & Goodwin, 2006). Microbial biosynthesis of esters can be catalyzed by either lipases (van den Berg, Heeres, van der Wielen, & Straathof, 2013) or alcohol acetyltransferases (Aat) (Layton & Trinh, 2014; Rodriguez et al., 2014). Lipases use the Fisher esterification chemistry, i.e. catalyze condensation of an alcohol and a carboxylic acid, while Aat catalyzes the reaction of an alcohol with an acyl-CoA. Hence, the AAT-dependent pathway is thermodynamically favorable, since it takes advantage from the high free energy stored in the thioester bond of acyl-CoA (Kruis et al., 2019; Layton & Trinh, 2014; Park et al., 2009; Rodriguez et al., 2014).

C. thermocellum has recently emerged as attractive producer of isobutyl acetate through fermentation of renewable biomass by taking advantage of its native ability to biosynthesize alcohols (e.g., ethanol and isobutanol) and acyl-CoAs (e.g., acetyl-CoA and isobutyryl-CoA) (Seo et al., 2020; Seo, Lee, Garcia, & Trinh, 2019). Isobutyl acetate is a biodegradable solvent which can replace more conventional alternatives (e.g., methyl isobutyl ketone or toluene) in many applications (Seo et al., 2020). So far, studies aimed at engineering isobutyl acetate production in *C. thermocellum* have been based on expression of a mutant heterologous thermostable chloramphenicol acetyltransferase (Cat) (Seo et al., 2019) and the deletion of two autologous carbohydrate esterases (Seo et al., 2020). Although the native function of the Cat protein is to transfer an acetyl group from acetyl-CoA to chloramphenicol, a protein synthesis inhibitor, this enzyme generally recognizes a broad range of alcohols and acyl-CoAs (Rodriguez et al., 2014). A thermotolerant mutant Cat from *Staphylococcus aureus* with increased affinity for isobutanol was expressed in *C. thermocellum* thus improving its production of isobutyl acetate and other esters (ethyl acetate, ethyl isobutyrate, and isobutyl isobutyrate) through cellulose fermentation (Seo et al., 2019). However, only 1.9 mg/L of isobutyl acetate was produced after 48 h. A number of hypotheses were raised for explaining this low efficiency including low affinity of the mutant Cat for isobutanol or its low expression. Additionally, *C. thermocellum* produces several carbohydrate esterases which are involved in plant biomass breakdown but could also be involved in other ester hydrolysis (such as isobutanol acetate), thus lowering ester production in this strain (Seo et al., 2020). Two carbohydrate esterases encoded by Clo1313_0613 and Clo1313_0693 have been shown to significantly contribute to the degradation of isobutyl acetate in *C. thermocellum*. These genes were therefore deleted resulting in improved isobutyl acetate production while maintaining effective cellulose depolymerization (Seo et al., 2020). Nonetheless, the maximum isobutyl acetate titer obtained so far in *C. thermocellum* is very low (3.1 mg/L after 72 h). For comparison, an *E. coli* strain expressing a *S. cerevisiae*-derived Aat (ATF1) was able to produce 17.5 g/L isobutyl acetate corresponding to 80% of the maximum theoretical yield (Rodriguez et al., 2014). One possible explanation is that the *S. cerevisiae* ATF1 has high affinity for isobutanol, while the *S. aureus*-derived Cat does not. The lack of a thermostable Aat represents an important barrier toward further improvement of isobutyl acetate and other esters in *C. thermocellum* (Seo et al., 2020). Protein engineering is likely necessary to address both isobutanol affinity and thermostability.

5. Conclusion

Ten years ago, the first example of metabolic engineering of *C. thermocellum* was reported. Since then, there has been impressive progress in understanding its metabolism and improving production of biofuels and other high-value chemicals through direct fermentation of cellulose (Table 1). The native ability of *C. thermocellum* to efficiently and rapidly solubilize cellulose is one of its major advantages. Gene manipulation tools developed so far proved to be robust and reliable and new tools have been recently developed thus expanding the possibilities of gene modification and more refined modulation of gene expression (Ganguly et al., 2020; Walker et al., 2020). Characterization of central sugar metabolism of this strain has highlighted some peculiar features (e.g., preference of co-factors, allosteric regulation, and peculiar metabolism of PEP) that likely influence the global metabolic network. An important signature of *C. thermocellum* glycolysis is that it works at nearly thermodynamic equilibrium (Jacobson et al., 2020). From an industrial viewpoint this is an issue, because this kind of metabolism may be more subject to inhibition by product feedback (Jacobson et al., 2020). Some attempts to improve *C. thermocellum* glycolytic flux (e.g., by introducing heterologous Gapdh or Pyk) have shown that this approach could help development of strains that are more suitable for industrial application (Deng et al., 2013; Tian, Perot, Stevenson, et al., 2017). Another important issue for *C. thermocellum* application to CBP of lignocellulose is its inability to use most other polysaccharide fractions of plant biomass, namely hemicellulose and pectin. So far, little progress has been made on this aspect. An engineered *C. thermocellum* strain expressing heterologous xylose isomerase and xylulose kinase was able to co-ferment xylose (and short xylo-oligosaccharides to a limited extent) and crystalline cellulose without carbon catabolite repression (Xiong, Reyes, et al., 2018). However, co-culture with thermophilic hemicellulolytic microorganisms has been shown to be an effective alternative for solubilizing and fermenting most of the sugars contained in different kinds of lignocellulosic biomass (e.g., corn fiber, wheat straw) (Beri et al., 2020; Froese et al., 2019).

Despite recent progress, attaining the yield, titer and productivity necessary for economic viability requires additional work. The greatest effort to date has been dedicated to increasing ethanol production. In terms of ethanol yield, the best strain obtained so far showed a value of 75% of the theoretical maximum and titer of 27 g/L from cellulose (Table 1) (Tian, Papanek, et al., 2016). This strain was obtained by disruption of the pathways for all the organic acids and hydrogen, followed by adaptive evolution. However, a similar titer and a faster growth (with minor yield reduction) was later obtained through introduction of genes from the pyruvate to ethanol pathway of *T. saccharolyticum* (Table 1) (Hon et al., 2018). Attempts to combine all these modifications, namely deleting pathways for competing end-products and introducing the *T. saccharolyticum* pyruvate to ethanol pathways, in a single strain were unsuccessful or have led to only modest increases in ethanol production (Hon et al., 2017). Further improvement of ethanol production seems limited by accumulation of glucose or lack of substrate consumption. Currently, a main metabolic bottleneck seems to be the PFK reaction (Cui et al., 2020; Dash et al., 2019; Jacobson et al., 2020). As research progresses, more detailed understanding of *C. thermocellum* metabolism and availability of new heterologous pathway genes will likely offer an expanded panel of strategies and tools to improve ethanol production in this strain. For instance, the limitations of genetic tools have made it difficult to transfer point mutations from one *C. thermocellum* strain to another for many years, but the very recent development of CRISPR-based genome editing tools for this organism may end this restriction (Walker et al., 2020). It is evident that enhancing production of ethanol in *C. thermocellum* is much more complicated than in organisms such as *S. cerevisiae* or engineered strains of *E. coli*. However, it is worth remembering that attempts to engineer a cellulolytic phenotype in *S. cerevisiae* have been even more hampered by the difficulty of secreting high amounts of heterologous cellulose-depolymerizing enzymes. As far as we know, the best-performing recombinant cellulolytic *S. cerevisiae* strain obtained to date was able to produce 3.8 g/L ethanol from crystalline cellulose with 75% yield in 96 h (Song et al., 2018). The best engineered *C. thermocellum* strains developed so far achieve higher performance in terms of both titer and productivity (Table 1).

Compared to the work done on ethanol production, studies aimed at increasing *n*-butanol and isobutanol production of *C. thermocellum* can be considered in their infancy. Of the two, more progress (in terms of yield and titer) has been obtained for isobutanol production. In fact, isobutanol is natively produced by *C. thermocellum* which has likely helped metabolic engineering strategies aimed at improving its production. *n*-Butanol titers obtained by engineered *C. thermocellum* are by far lower than those of isobutanol. Of all of the reports of cellulosic isobutanol production to date, *C. thermocellum* has shown the highest titer. For cellulosic *n*-butanol production, the reports of highest titer have come from *C. cellulovorans*. However, the number of options for future improvement of *n*-butanol and isobutanol production in *C. thermocellum* seems high and could also take advantage of much of the previous work aimed at improving ethanol production. A logical step would be to eliminate the native pathways that compete for carbon intermediates (e.g., acetate production) and/or reducing equivalents (e.g., ethanol production).

Today, lactic acid production through direct lignocellulose fermentation by *C. thermocellum* is more a wish than a reality. In general, little research effort has been dedicated at improving lactic acid production in cellulolytic microorganisms, so far, since most attention has been targeted to biofuels. This is certainly based on a larger economic interest in cellulosic biofuels (Lynd, 2017). However, since the spread of traditional (i.e., not biodegradable) plastics in almost every ecosystems on the Earth is an urgent environmental threat (Akdogan & Guven, 2019), incentives have increased for the development of bio-based biodegradable plastics such as PLA. Production of cellulosic lactic acid by using native cellulolytic microorganisms such as *C. thermocellum* is an attractive alternative to strategies based on engineering cellulolytic phenotype in lactic acid bacteria or other strong native producers of lactic acid. To date, the latter have been severely limited by issues in expressing heterologous cellulases (Mazzoli, 2020; Tarraran & Mazzoli, 2018). To the best of our knowledge, the highest performing strain obtained by this approach was able to directly ferment cello-oligosaccharides up to cellooctaose to L-LA at high yield (Gandini, Tarraran, Kalemasi, Pessione, & Mazzoli, 2017), but no engineered strain able to grow on more complex cellulosic substrates has been reported, so far. Notwithstanding the current gaps concerning lactate production in *C. thermocellum*, much work can be performed on this strain to transform its current potential for lactic acid production into a strain suitable for industrial application. Better understanding of the factors that contribute to modulate Ldh expression and activity and disruption of competing pathways are among the logical steps for progress in this direction.

Very recently, *C. thermocellum* has attracted interest also as producer of medium-chain esters from renewable biomass (Seo et al., 2019, 2020). So far, progress in this direction has been modest based on the low final titers obtained. It is worth remembering that these studies were the first attempts to engineer ester production in a thermophilic strain, which dramatically reduced the number of pathway genes available. Owing to the great potential of medium-chain esters as drop-in fuels, it is desirable that this line of research continue and that newly discovered genetic tools will likely aid its progress.

Acknowledgments

R.M. was supported by an Italy-U.S. Fulbright Research Scholarship. D.O. was supported by the Center for Bioenergy Innovation, a U.S. Department of Energy Research Center supported by the Office of Biological and Environmental Research in the DOE Office of Science.

Conflict of interest

Authors declare no conflict of interest.

References

Abdel-Rahman, M. A., & Sonomoto, K. (2016). Opportunities to overcome the current limitations and challenges for efficient microbial production of optically pure lactic acid. *Journal of Biotechnology*. https://doi.org/10.1016/j.jbiotec.2016.08.008.

Abdel-Rahman, M. A., Tashiro, Y., & Sonomoto, K. (2013). Recent advances in lactic acid production by microbial fermentation processes. *Biotechnology Advances, 31*, 877–902. https://doi.org/10.1016/j.biotechadv.2013.04.002.

Aburaya, S., Esaka, K., Morisaka, H., Kuroda, K., & Ueda, M. (2015). Elucidation of the recognition mechanisms for hemicellulose and pectin in *Clostridium cellulovorans* using intracellular quantitative proteome analysis. *AMB Express, 5*. https://doi.org/10.1186/s13568-015-0115-6.

Akdogan, Z., & Guven, B. (2019). Microplastics in the environment: A critical review of current understanding and identification of future research needs. *Environmental Pollution, 254*, 113011. https://doi.org/10.1016/j.envpol.2019.113011.

Akinosho, H., Yee, K., Close, D., & Ragauskas, A. (2014). The emergence of *Clostridium thermocellum* as a high utility candidate for consolidated bioprocessing applications. *Frontiers in Chemistry*. https://doi.org/10.3389/fchem.2014.00066.

Alves de Oliveira, R., Komesu, A., Vaz Rossell, C. E., & Maciel Filho, R. (2018). Challenges and opportunities in lactic acid bioprocess design—From economic to production aspects. *Biochemical Engineering Journal, 133*, 219–239. https://doi.org/10.1016/j.bej.2018.03.003.

Argyros, A., Barrett, T., Caiazza, N., & Hogsett, D. (2011). *Genetically modified* Clostridium thermocellum *engineered to ferment xylose*. CA2822654A1.

Argyros, D. A., Tripathi, S. A., Barrett, T. F., Rogers, S. R., Feinberg, L. F., Olson, D. G., et al. (2011). High ethanol Titers from cellulose by using metabolically engineered thermophilic, anaerobic microbes. *Applied and Environmental Microbiology, 77*, 8288–8294. https://doi.org/10.1128/AEM.00646-11.

Atsumi, S., Hanai, T., & Liao, J. C. (2008). Non-fermentative pathways for synthesis of branched-chain higher alcohols as biofuels. *Nature, 451*, 86–89. https://doi.org/10.1038/nature06450.

Atsumi, S., Higashide, W., & Liao, J. C. (2009). Direct photosynthetic recycling of carbon dioxide to isobutyraldehyde. *Nature Biotechnology, 27*, 1177–1180. https://doi.org/10.1038/nbt.1586.

Balch, M. L., Holwerda, E. K., Davis, M. F., Sykes, R. W., Happs, R. M., Kumar, R., et al. (2017). Lignocellulose fermentation and residual solids characterization for senescent switchgrass fermentation by: *Clostridium thermocellum* in the presence and absence of continuous in situ ball-milling. *Energy & Environmental Science, 10*, 1252–1261. https://doi.org/10.1039/c6ee03748h.

Bao, T., Zhao, J., Li, J., Liu, X., & Yang, S. T. (2019). n-Butanol and ethanol production from cellulose by *Clostridium cellulovorans* overexpressing heterologous aldehyde/alcohol dehydrogenases. *Bioresource Technology, 285*, 121316. https://doi.org/10.1016/j.biortech.2019.121316.

Bayer, E. A., Kenig, R., & Lamed, R. (1983). Adherence of *Clostridium thermocellum* to cellulose. *Journal of Bacteriology, 156*, 818–827. https://doi.org/10.1128/jb.156.2.818-827.1983.

Berezina, O. V., Zakharova, N. V., Brandt, A., Yarotsky, S. V., Schwarz, W. H., & Zverlov, V. V. (2010). Reconstructing the clostridial n-butanol metabolic pathway in *Lactobacillus brevis*. *Applied Microbiology and Biotechnology, 87*, 635–646. https://doi.org/10.1007/s00253-010-2480-z.

Beri, D., York, W. S., Lynd, L. R., Peña, M. J., & Herring, C. D. (2020). Development of a thermophilic coculture for corn fiber conversion to ethanol. *Nature Communications, 11*. https://doi.org/10.1038/s41467-020-15704-z.

Biddy, M. J., Scarlata, C. J., & Kinchin, C. M. (2016). Chemicals from biomass: A market assessment of bioproducts with near-term potential. *NREL Report*. https://doi.org/10.2172/1244312.

Biegel, E., Schmidt, S., González, J. M., & Müller, V. (2011). Biochemistry, evolution and physiological function of the Rnf complex, a novel ion-motive electron transport complex in prokaryotes. *Cellular and Molecular Life Sciences*. https://doi.org/10.1007/s00018-010-0555-8.

Biswas, R., Prabhu, S., Lynd, L. R., & Guss, A. M. (2014). Increase in ethanol yield via elimination of lactate production in an ethanol-tolerant mutant of *Clostridium thermocellum*. *PLoS One, 9*. https://doi.org/10.1371/journal.pone.0086389.

Biswas, R., Zheng, T., Olson, D. G., Lynd, L. R., & Guss, A. M. (2015). Elimination of hydrogenase active site assembly blocks H2 production and increases ethanol yield in *Clostridium thermocellum*. *Biotechnology for Biofuels, 8*, 20. https://doi.org/10.1186/s13068-015-0204-4.

Brown, S. D., Guss, A. M., Karpinets, T. V., Parks, J. M., Smolin, N., Yang, S., et al. (2011). Mutant alcohol dehydrogenase leads to improved ethanol tolerance in *Clostridium thermocellum*. *Proceedings of the National Academy of Sciences of the United States of America, 108*, 13752–13757. https://doi.org/10.1073/pnas.1102444108.

Bryant, F. O. (1991). Characterization of the fructose 1,6-bisphosphate-activated, L(+)lactate dehydrogenase from thermoanaerobacter ethanolicus. *Journal of Enzyme Inhibition and Medicinal Chemistry, 5*, 235–248. https://doi.org/10.3109/14756369109080062.

Buckel, W., & Thauer, R. K. (2013). Energy conservation via electron bifurcating ferredoxin reduction and proton/Na(+) translocating ferredoxin oxidation. *Biochimica et Biophysica Acta, 1827*, 94–113. https://doi.org/10.1016/j.bbabio.2012.07.002.

Campos-Fernández, J., Arnal, J. M., Gómez, J., & Dorado, M. P. (2012). A comparison of performance of higher alcohols/diesel fuel blends in a diesel engine. *Applied Energy, 95*, 267–275. https://doi.org/10.1016/j.apenergy.2012.02.051.

Chen, C. T., & Liao, J. C. (2016). Frontiers in microbial 1-butanol and isobutanol production. *FEMS Microbiology Letters*. https://doi.org/10.1093/femsle/fnw020.

Cui, J., Olson, D. G., & Lynd, L. R. (2019). Characterization of the *Clostridium thermocellum* AdhE, NfnAB, ferredoxin and Pfor proteins for their ability to support high titer ethanol production in Thermoanaerobacterium saccharolyticum. *Metabolic Engineering, 51*, 32–42. https://doi.org/10.1016/j.ymben.2018.09.006.

Cui, J., Stevenson, D., Korosh, T., Amador-noguez, D., Olson, D. G., Lynd, L. R., et al. (2020). Developing a Cell-Free Extract Reaction (CFER) system in *Clostridium thermocellum* to identify metabolic limitations to ethanol production. *Frontiers in Energy Research*. https://doi.org/10.3389/fenrg.2020.00072.

Dash, S., Olson, D. G., Joshua Chan, S. H., Amador-Noguez, D., Lynd, L. R., & Maranas, C. D. (2019). Thermodynamic analysis of the pathway for ethanol production from cellobiose in *Clostridium thermocellum*. *Metabolic Engineering, 55*, 161–169. https://doi.org/10.1016/j.ymben.2019.06.006.

Demain, A. L., Newcomb, M., & Wu, J. H. D. (2005). Cellulase, clostridia, and ethanol. *Microbiology and Molecular Biology Reviews, 69*, 124–154. https://doi.org/10.1128/mmbr.69.1.124-154.2005.

Deng, Y., Olson, D. G., Zhou, J., Herring, C. D., Joe Shaw, A., & Lynd, L. R. (2013). Redirecting carbon flux through exogenous pyruvate kinase to achieve high ethanol yields in *Clostridium thermocellum*. *Metabolic Engineering, 15*, 151–158. https://doi.org/10.1016/j.ymben.2012.11.006.

Deng, Y., Olson, D. G., Zhou, J., Herring, C. D., Joe Shaw, A., & Lynd, L. R. (2014). Redirecting carbon flux through exogenous pyruvate kinase to achieve high ethanol yields in *Clostridium thermocellum*. *Metabolic Engineering, 22*, 1–2. https://doi.org/10.1016/j.ymben.2013.11.006.

Desvaux, M., Guedon, E., & Petitdemange, H. (2000). Cellulose catabolism by *Clostridium cellulolyticum* growing in batch culture on defined medium. *Applied and Environmental Microbiology, 66*, 2461–2470. https://doi.org/10.1128/AEM.66.6.2461-2470.2000.

Dien, B. S., Cotta, M. A., & Jeffries, T. W. (2003). Bacteria engineered for fuel ethanol production: Current status. *Applied Microbiology and Biotechnology*. https://doi.org/10.1007/s00253-003-1444-y.

Eminoğlu, A., Murphy, S. J. L., Maloney, M., Lanahan, A., Giannone, R. J., Hettich, R. L., et al. (2017). Deletion of the hfsB gene increases ethanol production in Thermoanaerobacterium saccharolyticum and several other thermophilic anaerobic bacteria. *Biotechnology for Biofuels, 10*, 282. https://doi.org/10.1186/s13068-017-0968-9.

Ethanol Industry Outlook. (2019). *Powered with renewed energy*.

Feinberg, L., Foden, J., Barrett, T., Davenport, K. W., Bruce, D., Detter, C., et al. (2011). Complete genome sequence of the cellulolytic thermophile *Clostridium thermocellum* DSM1313. *Journal of Bacteriology*. https://doi.org/10.1128/JB.00322-11.

Froese, A., Schellenberg, J., & Sparling, R. (2019). Enhanced depolymerization and utilization of raw lignocellulosic material by co-cultures of Ruminiclostridium thermocellum with hemicellulose-utilizing partners. *Canadian Journal of Microbiology, 65*, 296–307. https://doi.org/10.1139/cjm-2018-0535.

Fuchs, K. P., Zverlov, V. V., Velikodvorskaya, G. A., Lottspeich, F., & Schwarz, W. H. (2003). Lic16A of Clostridium thermocellum, a non-cellulosomal, highly complex endo-β-1,3-glucanase bound to the outer cell surface. *Microbiology*. https://doi.org/10.1099/mic.0.26153-0.

Gaida, S. M., Liedtke, A., Jentges, A. H. W., Engels, B., & Jennewein, S. (2016). Metabolic engineering of *Clostridium cellulolyticum* for the production of n-butanol from crystalline cellulose. *Microbial Cell Factories15*. https://doi.org/10.1186/s12934-015-0406-2.

Gandini, C., Tarraran, L., Kalemasi, D., Pessione, E., & Mazzoli, R. (2017). Recombinant *Lactococcus lactis* for efficient conversion of cellodextrins into L-lactic acid. *Biotechnology and Bioengineering, 114*, 2807–2817. https://doi.org/10.1002/bit.26400.

Ganguly, J., Martin-Pascual, M., & van Kranenburg, R. (2020). CRISPR interference (CRISPRi) as transcriptional repression tool for *Hungateiclostridium thermocellum* DSM 1313. *Microbial Biotechnology, 13*, 339–349. https://doi.org/10.1111/1751-7915.13516.

Gray, K. A., Zhao, L., & Emptage, M. (2006). Bioethanol. *Current Opinion in Chemical Biology, 10*, 141–146.

Gu, J., Zhou, J., Zhang, Z., Kim, C. H., Jiang, B., Shi, J., et al. (2017). Isobutanol and 2-ketoisovalerate production by *Klebsiella pneumoniae* via a native pathway. *Metabolic Engineering, 43*, 71–84. https://doi.org/10.1016/j.ymben.2017.07.003.

Guedon, E., Desvaux, M., & Petitdemange, H. (2002). Improvement of cellulolytic properties of *Clostridium cellulolyticum* by metabolic engineering. *Applied and Environmental Microbiology, 68*, 53–58. https://doi.org/10.1128/AEM.68.1.53-58.2002.

Hamilton-Brehm, S. D., Mosher, J. J., Vishnivetskaya, T., Podar, M., Carroll, S., Allman, S., et al. (2010). Caldicellulosiruptor obsidiansis sp. nov., an anaerobic, extremely thermophilic, cellulolytic bacterium isolated from obsidian pool, Yellowstone National Park. *Applied and Environmental Microbiology, 76*, 1014–1020. https://doi.org/10.1128/AEM.01903-09.

Hannon, J. R., Lynd, L. R., Andrade, O., Benavides, P. T., Beckham, G. T., Biddy, M. J., et al. (2020). Technoeconomic and life-cycle analysis of single-step catalytic conversion of wet ethanol into fungible fuel blendstocks. *Proceedings of the National Academy of Sciences of the United States of America, 117*, 12576–12583. https://doi.org/10.1073/pnas.1821684116.

He, Q., Hemme, C. L., Jiang, H., He, Z., & Zhou, J. (2011). Mechanisms of enhanced cellulosic bioethanol fermentation by co-cultivation of *Clostridium* and *Thermoanaerobacter* spp. *Bioresource Technology, 102*, 9586–9592. https://doi.org/10.1016/j.biortech.2011.07.098.

Higashide, W., Li, Y., Yang, Y., & Liao, J. C. (2011). Metabolic engineering of *Clostridium cellulolyticum* for production of isobutanol from cellulose. *Applied and Environmental Microbiology, 77*, 2727–2733. https://doi.org/10.1128/AEM.02454-10.

Himmel, M. E., Ding, S. Y., Johnson, D. K., Adney, W. S., Nimlos, M. R., Brady, J. W., et al. (2007). Biomass recalcitrance: Engineering plants and enzymes for biofuels production. *Science*. https://doi.org/10.1126/science.1137016.

Holwerda, E. K., Olson, D. G., Ruppertsberger, N. M., Stevenson, D. M., Murphy, S. J. L., Maloney, M. I., et al. (2020). Metabolic and evolutionary responses of *Clostridium thermocellum* to genetic interventions aimed at improving ethanol production. *Biotechnology for Biofuels, 13*. https://doi.org/10.1186/s13068-020-01680-5.

Holwerda, E. K., Thorne, P. G., Olson, D. G., Amador-Noguez, D., Engle, N. L., Tschaplinski, T. J., et al. (2014). The exometabolome of *Clostridium thermocellum* reveals overflow metabolism at high cellulose loading. *Biotechnology for Biofuels, 7*. https://doi.org/10.1186/s13068-014-0155-1.

Holwerda, E. K., Worthen, R. S., Kothari, N., Lasky, R. C., Davison, B. H., Fu, C., et al. (2019). Multiple levers for overcoming the recalcitrance of lignocellulosic biomass. *Biotechnology for Biofuels, 12*. https://doi.org/10.1186/s13068-019-1353-7.

Hon, S., Holwerda, E. K., Worthen, R. S., Maloney, M. I., Tian, L., Cui, J., et al. (2018). Expressing the Thermoanaerobacterium saccharolyticum pforA in engineered *Clostridium thermocellum* improves ethanol production. *Biotechnology for Biofuels, 11*. https://doi.org/10.1186/s13068-018-1245-2.

Hon, S., Lanahan, A. A., Tian, L., Giannone, R. J., Hettich, R. L., Olson, D. G., et al. (2016). Development of a plasmid-based expression system in *Clostridium thermocellum* and its use to screen heterologous expression of bifunctional alcohol dehydrogenases (adhEs). *Metabolic Engineering Communications, 3*, 120–129. https://doi.org/10.1016/j.meteno.2016.04.001.

Hon, S., Olson, D. G., Holwerda, E. K., Lanahan, A. A., Murphy, S. J. L., Maloney, M. I., et al. (2017). The ethanol pathway from Thermoanaerobacterium saccharolyticum improves ethanol production in *Clostridium thermocellum*. *Metabolic Engineering, 42*, 175–184. https://doi.org/10.1016/j.ymben.2017.06.011.

Hu, L., Huang, H., Yuan, H., Tao, F., Xie, H., & Wang, S. (2016). Rex in Clostridium kluyveri is a global redox-sensing transcriptional regulator. *Journal of Biotechnology, 233*, 17–25. https://doi.org/10.1016/j.jbiotec.2016.06.024.

Ingram, L. O. (1976). Adaptation of membrane lipids to alcohols. *Journal of Bacteriology, 125*, 670–678. https://doi.org/10.1128/jb.125.2.670-678.1976.

Ishchuk, O. P., Voronovsky, A. Y., Stasyk, O. V., Gayda, G. Z., Gonchar, M. V., Abbas, C. A., et al. (2008). Overexpression of pyruvate decarboxylase in the yeast Hansenula polymorpha results in increased ethanol yield in high-temperature fermentation of xylose. *FEMS Yeast Research, 8*, 1164–1174. https://doi.org/10.1111/j.1567-1364.2008.00429.x.

Jacobson, T. B., Korosh, T. K., Stevenson, D. M., Foster, C., Maranas, C., Olson, D. G., et al. (2020). In vivo thermodynamic analysis of glycolysis in *Clostridium thermocellum* and Thermoanaerobacterium saccharolyticum using ^{13}C and ^{2}H tracers. *mSystems, 5*. https://doi.org/10.1128/msystems.00736-19.

Jiang, H. L., He, Q., He, Z., Hemme, C. L., Wu, L., & Zhou, J. (2013). Continuous cellulosic bioethanol fermentation by cyclic fed-batch cocultivation. *Applied and Environmental Microbiology, 79*, 1580–1589. https://doi.org/10.1128/AEM.02617-12.

Johns, N. I., Blazejewski, T., Gomes, A. L. C., & Wang, H. H. (2016). Principles for designing synthetic microbial communities. *Current Opinion in Microbiology*. https://doi.org/10.1016/j.mib.2016.03.010.

Kannuchamy, S., Mukund, N., & Saleena, L. M. (2016). Genetic engineering of *Clostridium thermocellum* DSM1313 for enhanced ethanol production. *BMC Biotechnology, 16*. https://doi.org/10.1186/s12896-016-0260-2.

Kata, I., Semkiv, M. V., Ruchala, J., Dmytruk, K. V., & Sibirny, A. A. (2016). Overexpression of the genes PDC1 and ADH1 activates glycerol conversion to ethanol in the thermotolerant yeast Ogataea (Hansenula) polymorpha. *Yeast, 33*, 471–478. https://doi.org/10.1002/yea.3175.

Kim, S. K., Groom, J., Chung, D., Elkins, J., & Westpheling, J. (2017). Expression of a heat-stable NADPH-dependent alcohol dehydrogenase from Thermoanaerobacter pseudethanolicus 39E in *Clostridium thermocellum* 1313 results in increased hydroxymethylfurfural resistance. *Biotechnology for Biofuels, 10*. https://doi.org/10.1186/s13068-017-0750-z.

Kim, S. K., & Westpheling, J. (2018). Engineering a spermidine biosynthetic pathway in *Clostridium thermocellum* results in increased resistance to furans and increased ethanol production. *Metabolic Engineering, 49*, 267–274. https://doi.org/10.1016/j.ymben.2018.09.002.

Kruis, A. J., Bohnenkamp, A. C., Patinios, C., van Nuland, Y. M., Levisson, M., Mars, A. E., et al. (2019). Microbial production of short and medium chain esters: Enzymes, pathways, and applications. *Biotechnology Advances.* https://doi.org/10.1016/j.biotechadv. 2019.06.006.

Lamed, R., Setter, E., & Bayer, E. A. (1983). Characterization of a cellulose-binding, cellulase-containing complex in clostridium thermocellum. *Journal of Bacteriology, 156,* 828–836. https://doi.org/10.1128/jb.156.2.828-836.1983.

Lamed, R., & Zeikus, J. G. (1981). Thermostable, ammonium-activated Malic Enzyme of *Clostridium thermocellum. Biochimica et Biophysica Acta, 660,* 251–255. https://doi.org/10.1016/0005-2744(81)90167-4.

Lange, J. P., Price, R., Ayoub, P. M., Louis, J., Petrus, L., Clarke, L., et al. (2010). Valeric biofuels: A platform of cellulosic transportation fuels. *Angewandte Chemie International Edition, 49,* 4479–4483. https://doi.org/10.1002/anie.201000655.

Layton, D. S., & Trinh, C. T. (2014). Engineering modular ester fermentative pathways in *Escherichia coli. Metabolic Engineering, 26,* 77–88. https://doi.org/10.1016/j.ymben.2014.09.006.

Legras, J. L., Merdinoglu, D., Cornuet, J. M., & Karst, F. (2007). Bread, beer and wine: *Saccharomyces cerevisiae* diversity reflects human history. *Molecular Ecology, 16,* 2091–2102. https://doi.org/10.1111/j.1365-294X.2007.03266.x.

Li, H., Opgenorth, P. H., Wernick, D. G., Rogers, S., Wu, T. Y., Higashide, W., et al. (2012). Integrated electromicrobial conversion of CO_2 to higher alcohols. *Science.* https://doi.org/10.1126/science.1217643.

Li, T., Zhang, C., Yang, K. L., & He, J. (2018). Unique genetic cassettes in a thermoanaerobacterium contribute to simultaneous conversion of cellulose and monosugars into butanol. *Science Advances.* https://doi.org/10.1126/sciadv.1701475.

Lian, J., Si, T., Nair, N. U., & Zhao, H. (2014). Design and construction of acetyl-CoA overproducing *Saccharomyces cerevisiae* strains. In *Food, pharmaceutical and bioengineering division 2014—Core programming area at the 2014 AIChE annual meeting* (pp. 750–760). https://doi.org/10.1016/j.ymben.2014.05.010.

Lin, P. P., Mi, L., Morioka, A. H., Yoshino, K. M., Konishi, S., Xu, S. C., et al. (2015). Consolidated bioprocessing of cellulose to isobutanol using *Clostridium thermocellum. Metabolic Engineering, 31,* 44–52. https://doi.org/10.1016/j.ymben.2015.07.001.

Lin, P. P., Rabe, K. S., Takasumi, J. L., Kadisch, M., Arnold, F. H., & Liao, J. C. (2014). Isobutanol production at elevated temperatures in thermophilic *Geobacillus thermoglucosidasius. Metabolic Engineering, 24,* 1–8. https://doi.org/10.1016/j.ymben.2014.03.006.

Linville, J. L., Rodriguez, M., Brown, S. D., Mielenz, J. R., & Cox, C. D. (2014). Transcriptomic analysis of *Clostridium thermocellum* Populus hydrolysate-tolerant mutant strain shows increased cellular efficiency in response to Populus hydrolysate compared to the wild type strain. *BMC Microbiology, 14.* https://doi.org/10.1186/s12866-014-0215-5.

Liu, Y., Lotero, E., & Goodwin, J. G. (2006). Effect of water on sulfuric acid catalyzed esterification. *Journal of Molecular Catalysis A: Chemical, 245,* 132–140. https://doi.org/10.1016/j.molcata.2005.09.049.

Lo, J., Olson, D. G., Murphy, S. J. L., Tian, L., Hon, S., Lanahan, A., et al. (2017). Engineering electron metabolism to increase ethanol production in *Clostridium thermocellum. Metabolic Engineering, 39,* 71–79. https://doi.org/10.1016/j.ymben.2016.10.018.

Lo, J., Zheng, T., Hon, S., Olson, D. G., & Lynd, L. R. (2015). The bifunctional alcohol and aldehyde dehydrogenase gene, adhE, is necessary for ethanol production in *Clostridium thermocellum* and Thermoanaerobacterium saccharolyticum. *Journal of Bacteriology, 197,* 1386–1393. https://doi.org/10.1128/JB.02450-14.

Lo, J., Zheng, T., Olson, D. G., Ruppertsberger, N., Tripathi, S. A., Guss, A. M., et al. (2015). Deletion of nfnAB in Thermoanaerobacterium saccharolyticum and its effect on metabolism. *Journal of Bacteriology, 197*, 2920–2929. https://doi.org/10.1128/JB.00347-15.

Lynd, L. R. (2017). The grand challenge of cellulosic biofuels. *Nature Biotechnology*. https://doi.org/10.1038/nbt.3976.

Lynd, L. R., Guss, A. M., Himmel, M. E., Beri, D., Herring, C., Holwerda, E. K., et al. (2016). Advances in consolidated bioprocessing using *Clostridium thermocellum* and *Thermoanaerobacter saccharolyticum*. In *Industrial biotechnology* (pp. 365–394). https://doi.org/10.1002/9783527807796.ch10.

Lynd, L. R., Liang, X., Biddy, M. J., Allee, A., Cai, H., Foust, T., et al. (2017). Cellulosic ethanol: Status and innovation. *Current Opinion in Biotechnology*. https://doi.org/10.1016/j.copbio.2017.03.008.

Lynd, L. R., Van Zyl, W. H., McBride, J. E., & Laser, M. (2005). Consolidated bioprocessing of cellulosic biomass: An update. *Current Opinion in Biotechnology*. https://doi.org/10.1016/j.copbio.2005.08.009.

Lynd, L. R., Weimer, P. J., van Zyl, W. H., & Pretorius, I. S. (2002). Microbial cellulose utilization: Fundamentals and biotechnology. *Microbiology and Molecular Biology Reviews, 66*, 739. https://doi.org/10.1128/mmbr.66.4.739.2002.

Marcano-Velazquez, J. G., Lo, J., Nag, A., Maness, P. C., & Chou, K. J. (2019). Developing riboswitch-mediated gene regulatory controls in thermophilic bacteria. *ACS Synthetic Biology, 8*, 633–640. https://doi.org/10.1021/acssynbio.8b00487.

Mazzoli, R. (2020). Metabolic engineering strategies for consolidated production of lactic acid from lignocellulosic biomass. *Biotechnology and Applied Biochemistry*. https://doi.org/10.1002/bab.1869.

Mazzoli, R., Bosco, F., Mizrahi, I., Bayer, E. A., & Pessione, E. (2014). Towards lactic acid bacteria-based biorefineries. *Biotechnology Advances*. https://doi.org/10.1016/j.biotechadv.2014.07.005.

McBee, R. H. (1954). The characteristics of *Clostridium thermocellum*. *Journal of Bacteriology, 67*, 505–506. https://doi.org/10.1128/jb.67.4.505-506.1954.

Mearls, E. B., Olson, D. G., Herring, C. D., & Lynd, L. R. (2015). Development of a regulatable plasmid-based gene expression system for *Clostridium thermocellum*. *Applied Microbiology and Biotechnology*. https://doi.org/10.1007/s00253-015-6610-5.

Morag, E., Bayer, E. A., & Lamed, R. (1990). Relationship of cellulosomal and non-cellulosomal xylanases of *Clostridium thermocellum* to cellulose-degrading enzymes. *Journal of Bacteriology, 172*, 6098–6105. https://doi.org/10.1128/jb.172.10.6098-6105.1990.

Nicolaou, S. A., Gaida, S. M., & Papoutsakis, E. T. (2010). A comparative view of metabolite and substrate stress and tolerance in microbial bioprocessing: From biofuels and chemicals, to biocatalysis and bioremediation. *Metabolic Engineering*. https://doi.org/10.1016/j.ymben.2010.03.004.

Nielsen, D. R., Leonard, E., Yoon, S. H., Tseng, H. C., Yuan, C., & Prather, K. L. J. (2009). Engineering alternative butanol production platforms in heterologous bacteria. *Metabolic Engineering, 11*, 262–273. https://doi.org/10.1016/j.ymben.2009.05.003.

Noor, E., Bar-Even, A., Flamholz, A., Reznik, E., Liebermeister, W., & Milo, R. (2014). Pathway thermodynamics highlights kinetic obstacles in central metabolism. *PLoS Computational Biology 10*. https://doi.org/10.1371/journal.pcbi.1003483.

Olson, D. G., Hörl, M., Fuhrer, T., Cui, J., Zhou, J., Maloney, M. I., et al. (2017). Glycolysis without pyruvate kinase in *Clostridium thermocellum*. *Metabolic Engineering*. https://doi.org/10.1016/j.ymben.2016.11.011.

Olson, D. G., & Lynd, L. R. (2012). Transformation of *Clostridium thermocellum* by electroporation. In *Methods in enzymology* (pp. 317–330). https://doi.org/10.1016/B978-0-12-415931-0.00017-3.

Olson, D. G., Maloney, M., Lanahan, A. A., Hon, S., Hauser, L. J., & Lynd, L. R. (2015). Identifying promoters for gene expression in *Clostridium thermocellum*. *Metabolic Engineering Communications, 2*, 23–29. https://doi.org/10.1016/j.meteno.2015.03.002.

Olson, D. G., Sparling, R., & Lynd, L. R. (2015). Ethanol production by engineered thermophiles. *Current Opinion in Biotechnology*. https://doi.org/10.1016/j.copbio.2015.02.006.

Özkan, M., Yilmaz, E. I., Lynd, L. R., & Özcengiz, G. (2004). Cloning and expression of the *Clostridium thermocellum* L-lactate dehydrogenase gene in *Escherichia coli* and enzyme characterization. *Canadian Journal of Microbiology, 50*, 845–851. https://doi.org/10.1139/w04-071.

Papanek, B., Biswas, R., Rydzak, T., & Guss, A. M. (2015). Elimination of metabolic pathways to all traditional fermentation products increases ethanol yields in *Clostridium thermocellum*. *Metabolic Engineering, 32*, 49–54. https://doi.org/10.1016/j.ymben.2015.09.002.

Park, Y. C., Shaffer, C. E. H., & Bennett, G. N. (2009). Microbial formation of esters. *Applied Microbiology and Biotechnology*. https://doi.org/10.1007/s00253-009-2170-x.

Raj, K. C., Talarico, L. A., Ingram, L. O., & Maupin-Furlow, J. A. (2002). Cloning and characterization of the *Zymobacter palmae* pyruvate decarboxylase gene (*pdc*) and comparison to bacterial homologues. *Applied and Environmental Microbiology, 68*, 2869–2876. https://doi.org/10.1128/AEM.68.6.2869.

Raman, B., Pan, C., Hurst, G. B., Rodriguez, M., McKeown, C. K., Lankford, P. K., et al. (2009). Impact of pretreated switchgrass and biomass carbohydrates on *Clostridium thermocellum* ATCC 27405 cellulosome composition: A quantitative proteomic analysis. *PLoS One4*. https://doi.org/10.1371/journal.pone.0005271.

Ravcheev, D. A., Li, X., Latif, H., Zengler, K., Leyn, S. A., Korostelev, Y. D., et al. (2012). Transcriptional regulation of central carbon and energy metabolism in bacteria by redox-responsive repressor rex. *Journal of Bacteriology, 194*, 1145–1157. https://doi.org/10.1128/JB.06412-11.

Riley, L. A., Ji, L., Schmitz, R. J., Westpheling, J., & Guss, A. M. (2019). Rational development of transformation in *Clostridium thermocellum* ATCC 27405 via complete methylome analysis and evasion of native restriction–modification systems. *Journal of Industrial Microbiology & Biotechnology, 46*, 1435–1443. https://doi.org/10.1007/s10295-019-02218-x.

Rodriguez, G. M., Tashiro, Y., & Atsumi, S. (2014). Expanding ester biosynthesis in *Escherichia coli*. *Nature Chemical Biology, 10*, 259–265. https://doi.org/10.1038/nchembio.1476.

Rydzak, T., García, D., Stevenson, D. M., Sladek, M., Klingeman, D. M., Holwerda, E. K., et al. (2017). Deletion of Type I glutamine synthetase deregulates nitrogen metabolism and increases ethanol production in *Clostridium thermocellum*. *Metabolic Engineering, 41*, 182–191. https://doi.org/10.1016/j.ymben.2017.04.002.

Rydzak, T., Lynd, L. R., & Guss, A. M. (2015). Elimination of formate production in *Clostridium thermocellum*. *Journal of Industrial Microbiology & Biotechnology, 42*, 1263–1272. https://doi.org/10.1007/s10295-015-1644-3.

Rydzak, T., McQueen, P. D., Krokhin, O. V., Spicer, V., Ezzati, P., Dwivedi, R. C., et al. (2012). Proteomic analysis of *Clostridium thermocellum* core metabolism: Relative protein expression profiles and growth phase-dependent changes in protein expression. *BMC Microbiology, 12*, 214. https://doi.org/10.1186/1471-2180-12-214.

Sander, K., Chung, D., Hyatt, D., Westpheling, J., Klingeman, D. M., Rodriguez, M., et al. (2019). Rex in *Caldicellulosiruptor bescii*: Novel regulon members and its effect on the production of ethanol and overflow metabolites. *Microbiology, 8*, e00639. https://doi.org/10.1002/mbo3.639.

Saxena, S., Fierobe, H. P., Gaudin, C., Guerlesquin, F., & Belaich, J. P. (1995). Biochemical properties of a β-xylosidase from *Clostridium cellulolyticum*. *Applied and Environmental Microbiology, 61*, 3509–3512. https://doi.org/10.1128/aem.61.9.3509-3512.1995.

Schwarz, K. M., Grosse-Honebrink, A., Derecka, K., Rotta, C., Zhang, Y., & Minton, N. P. (2017). Towards improved butanol production through targeted genetic modification of *Clostridium pasteurianum*. *Metabolic Engineering, 40*, 124–137. https://doi.org/10.1016/j.ymben.2017.01.009.

Seo, H., Lee, J. W., Garcia, S., & Trinh, C. T. (2019). Single mutation at a highly conserved region of chloramphenicol acetyltransferase enables isobutyl acetate production directly from cellulose by *Clostridium thermocellum* at elevated temperatures. *Biotechnology for Biofuels, 12*. https://doi.org/10.1186/s13068-019-1583-8.

Seo, H., Nicely, P. N., & Trinh, C. T. (2020). Endogenous carbohydrate esterases of *Clostridium thermocellum* are identified and disrupted for enhanced isobutyl acetate production from cellulose. *Biotechnology and Bioengineering*. https://doi.org/10.1002/bit.27360.

Shao, X. J., Raman, B., Zhu, M. J., Mielenz, J. R., Brown, S. D., Guss, A. M., et al. (2011). Mutant selection and phenotypic and genetic characterization of ethanol-tolerant strains of *Clostridium thermocellum*. *Applied Microbiology and Biotechnology, 92*, 641–652. https://doi.org/10.1007/s00253-011-3492-z.

Shen, C. R., Lan, E. I., Dekishima, Y., Baez, A., Cho, K. M., & Liao, J. C. (2011). Driving forces enable high-titer anaerobic 1-butanol synthesis in *Escherichia coli*. *Applied and Environmental Microbiology, 77*, 2905–2915. https://doi.org/10.1128/AEM.03034-10.

Shi, S., Si, T., Liu, Z., Zhang, H., Ang, E. L., & Zhao, H. (2016). Metabolic engineering of a synergistic pathway for n-butanol production in Saccharomyces cerevisiae. *Scientific Reports, 6*. https://doi.org/10.1038/srep25675.

Sleat, R., Mah, R. A., & Robinson, R. (1984). Isolation and characterization of an anaerobic, cellulolytic bacterium, Clostridium cellulovorans sp. nov. *Applied and Environmental Microbiology, 48*, 88–93. https://doi.org/10.1099/00207713-32-1-87.

Smith, K. M., Cho, K. M., & Liao, J. C. (2010). Engineering *Corynebacterium glutamicum* for isobutanol production. *Applied Microbiology and Biotechnology, 87*, 1045–1055. https://doi.org/10.1007/s00253-010-2522-6.

Song, X., Li, Y., Wu, Y., Cai, M., Liu, Q., Gao, K., et al. (2018). Metabolic engineering strategies for improvement of ethanol production in cellulolytic *Saccharomyces cerevisiae*. *FEMS Yeast Research, 18*. https://doi.org/10.1093/femsyr/foy090.

Steen, E. J., Chan, R., Prasad, N., Myers, S., Petzold, C. J., Redding, A., et al. (2008). Metabolic engineering of *Saccharomyces cerevisiae* for the production of n-butanol. *Microbial Cell Factories, 7*. https://doi.org/10.1186/1475-2859-7-36.

Taillefer, M., Rydzak, T., Levin, D. B., Oresnik, I. J., & Sparling, R. (2015). Reassessment of the transhydrogenase/malate shunt pathway in *Clostridium thermocellum* ATCC 27405 through kinetic characterization of malic enzyme and malate dehydrogenase. *Applied and Environmental Microbiology, 81*, 2423–2432. https://doi.org/10.1128/AEM.03360-14.

Tarraran, L., & Mazzoli, R. (2018). Alternative strategies for lignocellulose fermentation through lactic acid bacteria: The state of the art and perspectives. *FEMS Microbiology Letters*. https://doi.org/10.1093/femsle/fny126.

Tian, L., Cervenka, N. D., Low, A. M., Olson, D. G., & Lynd, L. R. (2019). A mutation in the AdhE alcohol dehydrogenase of *Clostridium thermocellum* increases tolerance to several primary alcohols, including isobutanol, n-butanol and ethanol. *Scientific Reports, 9*. https://doi.org/10.1038/s41598-018-37979-5.

Tian, L., Conway, P. M., Cervenka, N. D., Cui, J., Maloney, M., Olson, D. G., et al. (2019). Metabolic engineering of *Clostridium thermocellum* for n-butanol production from cellulose. *Biotechnology for Biofuels, 12*. https://doi.org/10.1186/s13068-019-1524-6.

Tian, L., Lo, J., Shao, X., Zheng, T., Olson, D. G., & Lynd, L. R. (2016). Ferredoxin: NAD + oxidoreductase of *Thermoanaerobacterium saccharolyticum* and its role in ethanol formation. *Applied and Environmental Microbiology, 82*, 7134–7141. https://doi.org/10.1128/AEM.02130-16.

Tian, L., Papanek, B., Olson, D. G., Rydzak, T., Holwerda, E. K., Zheng, T., et al. (2016). Simultaneous achievement of high ethanol yield and titer in *Clostridium thermocellum*. *Biotechnology for Biofuels*, 9, 116. https://doi.org/10.1186/s13068-016-0528-8.

Tian, L., Perot, S. J., Hon, S., Zhou, J., Liang, X., Bouvier, J. T., et al. (2017). Enhanced ethanol formation by *Clostridium thermocellum* via pyruvate decarboxylase. *Microbial Cell Factories*, 16. https://doi.org/10.1186/s12934-017-0783-9.

Tian, L., Perot, S. J., Stevenson, D., Jacobson, T., Lanahan, A. A., Amador-Noguez, D., et al. (2017). Metabolome analysis reveals a role for glyceraldehyde 3-phosphate dehydrogenase in the inhibition of C. thermocellum by ethanol. *Biotechnology for Biofuels*. https://doi.org/10.1186/s13068-017-0961-3.

Tomas, C. A., Welker, N. E., & Papoutsakis, E. T. (2003). Overexpression of groESL in *Clostridium acetobutylicum* results in increased solvent production and tolerance, prolonged metabolism, and changes in the cell's transcriptional program. *Applied and Environmental Microbiology*, 69, 4951–4965. https://doi.org/10.1128/AEM.69.8.4951-4965.2003.

Tripathi, S. A., Olson, D. G., Argyros, D. A., Miller, B. B., Barrett, T. F., Murphy, D. M., et al. (2010). Development of pyrF-based genetic system for targeted gene deletion in *Clostridium thermocellum* and creation of a pta mutant. *Applied and Environmental Microbiology*, 76, 6591–6599. https://doi.org/10.1128/AEM.01484-10.

Usai, G., Cirrincione, S., Re, A., Manfredi, M., Pagnani, A., Pessione, E., et al. (2020). Clostridium cellulovorans metabolism of cellulose as studied by comparative proteomic approach. *Journal of Proteomics, 216*. https://doi.org/10.1016/j.jprot.2020.103667.

van den Berg, C., Heeres, A. S., van der Wielen, L. A. M., & Straathof, A. J. J. (2013). Simultaneous clostridial fermentation, lipase-catalyzed esterification, and ester extraction to enrich diesel with butyl butyrate. *Biotechnology and Bioengineering*, 110, 137–142. https://doi.org/10.1002/bit.24618.

van der Veen, D., Lo, J., Brown, S. D., Johnson, C. M., Tschaplinski, T. J., Martin, M., et al. (2013). Characterization of *Clostridium thermocellum* strains with disrupted fermentation end-product pathways. *Journal of Industrial Microbiology & Biotechnology*, 40, 725–734. https://doi.org/10.1007/s10295-013-1275-5.

Van Zyl, L. J., Taylor, M. P., Eley, K., Tuffin, M., & Cowan, D. A. (2013). Engineering pyruvate decarboxylase-mediated ethanol production in the thermophilic host *Geobacillus thermoglucosidasius*. *Applied Microbiology and Biotechnology*. https://doi.org/10.1007/s00253-013-5380-1.

Vane, L. M. (2008). Separation technologies for the recovery and dehydration of alcohols from fermentation broths. *Biofuels, Bioproducts, Biorefining, 2*, 553–588. https://doi.org/10.1002/bbb.108.

Verbeke, T. J., Giannone, R. J., Klingeman, D. M., Engle, N. L., Rydzak, T., Guss, A. M., et al. (2017). Pentose sugars inhibit metabolism and increase expression of an AgrD-type cyclic pentapeptide in *Clostridium thermocellum*. *Scientific Reports, 7*. https://doi.org/10.1038/srep43355.

Walker, J. E., Lanahan, A. A., Zheng, T., Toruno, C., Lynd, L. R., Cameron, J. C., et al. (2020). Development of both type I–B and type II CRISPR/Cas genome editing systems in the cellulolytic bacterium *Clostridium thermocellum*. *Metabolic Engineering Communications, 10*. https://doi.org/10.1016/j.mec.2019.e00116.

Wang, M., Han, J., Dunn, J. B., Cai, H., & Elgowainy, A. (2015). Well-to-wheels energy use and greenhouse gas emissions of ethanol from corn, sugarcane and cellulosic biomass for US use. In *Efficiency and sustainability in biofuel production: Environmental and land-use research* (pp. 249–280). https://doi.org/10.1088/1748-9326/7/4/045905.

Wang, S., Huang, H., Moll, J., & Thauer, R. K. (2010). NADP+ reduction with reduced ferredoxin and NADP+ reduction with NADH are coupled via an electron-bifurcating enzyme complex in *Clostridium kluyveri*. *Journal of Bacteriology, 192*, 5115–5123. https://doi.org/10.1128/JB.00612-10.

Weimer, P. J., & Zeikus, J. G. (1977). Fermentation of cellulose and cellobiose by *Clostridium thermocellum* in the absence and presence of *Methanobacterium thermoautotrophicum*. *Applied and Environmental Microbiology, 33*, 289–297.

Wen, Z., Ledesma-Amaro, R., Lin, J., Jiang, Y., & Yangd, S. (2019). Improved n-butanol production from Clostridium cellulovorans by integrated metabolic and evolutionary engineering. *Applied and Environmental Microbiology, 85*. https://doi.org/10.1128/AEM.02560-18.

Wen, Z., Ledesma-Amaro, R., Lu, M., Jiang, Y., Gao, S., Jin, M., et al. (2020). Combined evolutionary engineering and genetic manipulation improve low pH tolerance and butanol production in a synthetic microbial Clostridium community. *Biotechnology and Bioengineering*. https://doi.org/10.1002/bit.27333.

Wen, Z., Ledesma-Amaro, R., Lu, M., Jin, M., & Yang, S. (2020). Metabolic engineering of Clostridium cellulovorans to improve butanol production by consolidated bioprocessing. *ACS Synthetic Biology*. https://doi.org/10.1021/acssynbio.9b00331.

Whitham, J. M., Moon, J. W., Rodriguez, M., Engle, N. L., Klingeman, D. M., Rydzak, T., et al. (2018). *Clostridium thermocellum* LL1210 pH homeostasis mechanisms informed by transcriptomics and metabolomics. *Biotechnology for Biofuels, 11*, 98. https://doi.org/10.1186/s13068-018-1095-y.

Wietzke, M., & Bahl, H. (2012). The redox-sensing protein Rex, a transcriptional regulator of solventogenesis in *Clostridium acetobutylicum*. *Applied Microbiology and Biotechnology, 96*, 749–761. https://doi.org/10.1007/s00253-012-4112-2.

Williams, T. I., Combs, J. C., Lynn, B. C., & Strobel, H. J. (2007). Proteomic profile changes in membranes of ethanol-tolerant *Clostridium thermocellum*. *Applied Microbiology and Biotechnology, 74*, 422–432. https://doi.org/10.1007/s00253-006-0689-7.

Williams-Rhaesa, A. M., Awuku, N. K., Lipscomb, G. L., Poole, F. L., Rubinstein, G. M., Conway, J. M., et al. (2018). Native xylose-inducible promoter expands the genetic tools for the biomass-degrading, extremely thermophilic bacterium *Caldicellulosiruptor bescii*. *Extremophiles, 22*, 629–638. https://doi.org/10.1007/s00792-018-1023-x.

Willquist, K., & van Niel, E. W. J. (2010). Lactate formation in *Caldicellulosiruptor saccharolyticus* is regulated by the energy carriers pyrophosphate and ATP. *Metabolic Engineering, 12*, 282–290. https://doi.org/10.1016/j.ymben.2010.01.001.

Wilson, C. M., Yang, S., Rodriguez, M., Ma, Q., Johnson, C. M., Dice, L., et al. (2013). Clostridium thermocellum transcriptomic profiles after exposure to furfural or heat stress. *Biotechnology for Biofuels, 6*. https://doi.org/10.1186/1754-6834-6-131.

Wu, C. W., Spike, T., Klingeman, D. M., Rodriguez, M., Bremer, V. R., & Brown, S. D. (2017). Generation and characterization of acid tolerant Fibrobacter succinogenes S85. *Scientific Reports, 7*, 2277. https://doi.org/10.1038/s41598-017-02628-w.

Xiong, W., Lin, P. P., Magnusson, L., Warner, L., Liao, J. C., Maness, P. C., et al. (2016). CO_2-fixing one-carbon metabolism in a cellulose-degrading bacterium *Clostridium thermocellum*. *Proceedings of the National Academy of Sciences of the United States of America, 113*, 13180–13185. https://doi.org/10.1073/pnas.1605482113.

Xiong, W., Lo, J., Chou, K. J., Wu, C., Magnusson, L., Dong, T., et al. (2018). Isotope-assisted metabolite analysis sheds light on central carbon metabolism of a model cellulolytic bacterium *Clostridium thermocellum*. *Frontiers in Microbiology, 9*. https://doi.org/10.3389/fmicb.2018.01947.

Xiong, W., Reyes, L. H., Michener, W. E., Maness, P. C., & Chou, K. J. (2018). Engineering cellulolytic bacterium *Clostridium thermocellum* to co-ferment cellulose- and hemicellulose-derived sugars simultaneously. *Biotechnology and Bioengineering, 115*, 1755–1763. https://doi.org/10.1002/bit.26590.

Yang, X., Xu, M., & Yang, S. T. (2015). Metabolic and process engineering of *Clostridium cellulovorans* for biofuel production from cellulose. *Metabolic Engineering, 32*, 39–48. https://doi.org/10.1016/j.ymben.2015.09.001.

Zheng, T., Lanahan, A. A., Lynd, L. R., & Olson, D. G. (2018). The redox-sensing protein Rex modulates ethanol production in *Thermoanaerobacterium saccharolyticum*. *PLoS One, 13*, e0195143. https://doi.org/10.1371/journal.pone.0195143.

Zheng, T., Olson, D. G., Murphy, S. J., Shao, X., Tian, L., & Lynd, L. R. (2017). Both adhE and a separate NADPH-dependent alcohol dehydrogenase gene, *adhA*, are necessary for high ethanol production in *Thermoanaerobacterium saccharolyticum*. *Journal of Bacteriology, 199*, e00542-16. https://doi.org/10.1128/JB.00542-16.

Zheng, T., Olson, D. G., Tian, L., Bomble, Y. J., Himmel, M. E., Lo, J., et al. (2015). Cofactor specificity of the bifunctional alcohol and aldehyde dehydrogenase (AdhE) in wild-type and mutant *Clostridium thermocellum* and thermoanaerobacterium saccharolyticum. *Journal of Bacteriology, 197*, 2610–2619. https://doi.org/10.1128/JB.00232-15.

Zhou, J., Olson, D. G., Argyros, D. A., Deng, Y., van Gulik, W. M., van Dijken, J. P., et al. (2013). Atypical glycolysis in *Clostridium thermocellum*. *Applied and Environmental Microbiology*. https://doi.org/10.1128/AEM.04037-12.

Zhou, J., Olson, D. G., Lanahan, A. A., Tian, L., Murphy, S. J. L., Lo, J., et al. (2015). Physiological roles of pyruvate ferredoxin oxidoreductase and pyruvate formate-lyase in *Thermoanaerobacterium saccharolyticum* JW/SL-YS485. *Biotechnology for Biofuels, 8*, 138. https://doi.org/10.1186/s13068-015-0304-1.

CHAPTER FOUR

Predetermined clockwork microbial worlds: Current understanding of aquatic microbial diel response from model systems to complex environments

Daichi Morimoto[a,*], Sigitas Šulčius[b], Kento Tominaga[a], and Takashi Yoshida[a,*]

[a]Graduate School of Agriculture, Kyoto University, Kyoto, Japan
[b]Laboratory of Algology and Microbial Ecology, Nature Research Centre, Vilnius, Lithuania
*Corresponding author: e-mail address: yoshidaten@gmail.com

Contents

1. Introduction 164
2. Culture-based understanding of microbial diel patterns at each trophic level 167
   2.1 Photoautotrophic microorganisms 167
   2.2 Heterotrophic/mixotrophic microorganisms 169
   2.3 Viruses 171
3. Community diel cycles generate from complex biological interactions 173
   3.1 Tracking complex interactions within microbial communities 173
   3.2 Diel cycling of microbial communities generated by direct interactions between phototrophic and heterotrophic microorganisms 174
   3.3 Diel cycling of microbial communities generated by indirect interactions 177
4. Conclusions and perspectives 180
Acknowledgments 182
References 182

Abstract

In the photic zone of aquatic ecosystems, microorganisms with different metabolisms and their viruses form complex interactions and food webs. Within these interactions, phototrophic microorganisms such as eukaryotic microalgae and cyanobacteria interact directly with sunlight, and thereby generate circadian rhythms. Diel cycling originally generated in microbial phototrophs is directly transmitted toward heterotrophic microorganisms utilizing the photosynthetic products as they are excreted or exuded. Such diel cycling seems to be indirectly propagated toward heterotrophs as a result of complex biotic interactions. For example, cell death of phototrophic microorganisms induced by viral lysis and protistan grazing provides additional resources of dissolved organic matter to the microbial community, and so generates diel cycling in other heterotrophs with different nutrient dependencies. Likewise, differences in the diel transmitting pathway via complex interactions among heterotrophs, and between heterotrophs and their viruses, may also generate higher variation and time lag diel rhythms in different heterotrophic taxa. Thus, sunlight and photosynthesis not only contribute energy and carbon supply, but also directly or indirectly control diel cycling of the microbial community through complex interactions in the photic zone of aquatic ecosystems.

1. Introduction

The marine environment covers 70% of Earth's surface. Prokaryotic and eukaryotic microbes represent ∼60% of ocean biomass and are key players in the surface marine food web (Bar-On, Phillips, & Milo, 2018). In the photic zone, phytoplankton (including cyanobacteria and eukaryotic microalgae such as diatoms and coccolithophorids) are major contributors to primary production, responsible for approximately 50% (50 GtC/year) of Earth's primary production (Field, Behrenfeld, Randerson, & Falkowski, 1998). The majority (at least half) of oceanic fixed carbon released from phytoplankton as dissolved organic matter (DOM) is remineralized by heterotrophic prokaryotes, mostly bacteria, which have been estimated to represent 10^{29} cells (Whitman, Coleman, & Wiebe, 1998). Subsequently, the carbon utilized by prokaryotes is channeled back to the classic food web through predation by microbial eukaryotes such as ciliates and heterotrophic dinoflagellates (Azam et al., 1983; Buchan, LeCleir, Gulvik, & González, 2014; Field et al., 1998; Worden et al., 2015). This flux is the so-called microbial loop and is considered to be the major pathway of marine carbon flux (Azam et al., 1983).

Viruses outnumber microorganisms in the aquatic ecosystems. Microorganisms are the hosts of most viruses, and so viruses are thought to play an important role in the rapid decay of their microbial host cells and release of cellular constituents to the pools of DOM (Breitbart, Bonnain, Malki, & Sawaya, 2018; Roux et al., 2016; Suttle, 2005, 2007). This prevailing concept in the biogeochemical cycle of marine viruses has been called the "viral shunt" (Suttle, 2005, 2007). In addition, viruses alter host metabolism during infection, through the expression of viral genes known as auxiliary metabolic genes (AMGs) (Breitbart, Thompson, Suttle, & Sullivan,

2007; Hurwitz & U'Ren, 2016). Recent estimation of the relative contribution of cyanoviral AMGs in cyanobacterial Photosystem II *psbA* suggests that, occasionally, >50% of expression is viral in origin, highlighting the large contribution of viruses to photosynthesis and oxygen production in the marine environment (Sieradzki, Ignacio-Espinoza, Needham, Fichot, & Fuhrman, 2019).

Microbial communities change over multiple timescales in response to environmental variables (Fuhrman, Cram, & Needham, 2015). However, as the typical average generation times of marine plankton are approximately 1 day in surface waters (Ducklow, 2000), changes in community composition at relatively short timescales have not previously been expected. Therefore, monthly sampling has been the most common interval in major ocean time-series studies (Fuhrman et al., 2015) and seasonal patterns in microbial communities have been the most thoroughly studied (reviewed in Bunse & Pinhassi, 2017; Faust, Lahti, Gonze, de Vos, & Raes, 2015; Fuhrman et al., 2015). Changes in the solar angle result in various cyclical seasonal changes (such as temperature, day length, weather patterns, land runoff, atmospheric deposition and upwelling associated with changes in nutrient availability) and affect microbial composition (Fuhrman et al., 2015). For example, the monthly bacterial composition pattern is reported as being repeatable over several years, and it has been suggested that seasonal environmental factors can predict this temporal pattern (Fuhrman et al., 2006). This strong seasonality and reoccurrence were observed in the prokaryotic community throughout the surface-to-bottom water column (Cram et al., 2015), and also observed in the marine viral community (Ignacio-Espinoza, Ahlgren, & Fuhrman, 2020; Pagarete et al., 2013) and in eukaryotes (Collado-Fabbri, Vaulot, & Ulloa, 2011; Marquardt, Vader, Stübner, Reigstad, & Gabrielsen, 2016).

Recent high-frequency sampling schemes (e.g., daily to weekly) have given extensive insights into the more fine-scale temporal changes in microbial communities. These studies revealed the appearance of sharp, major short-term peaks (within few days) in abundance of a particular population of microbial organism which responds to strong and ephemeral pulses in (in)organic matters (e.g., spring algal blooms) and dynamic succession of these abundant populations according to the change in available environmental substrate (Alonso-Sáez, Díaz-Pérez, & Morán, 2015; Lindh et al., 2015; Needham et al., 2013; Needham & Fuhrman, 2016; Teeling et al., 2012, 2016). Importantly, these temporally abundant (∼30% of total prokaryotic relative abundance) taxa include not only previously uncultured linages such as uncultured marine flavobacteria (e.g., the NS4 marine group) but also Marine groups II euryarchaeota, which were not traditionally thought to be rapidly growing organisms (Needham et al., 2013; Needham & Fuhrman, 2016; Teeling et al., 2012, 2016). Furthermore, these temporally abundant phytoplankton and prokaryotic dynamics correlate better with each other than with environmental parameters, suggesting that these temporal dynamics are majorly controlled by interactions among microorganisms (Needham & Fuhrman, 2016).

At a shorter, high-resolution time scale, diel biological responses, physiological and genetical responses have received much attention from the chronobiological aspect in not only microorganisms but also animals and plants (Bhadra, Thakkar, Das, & Pal Bhadra, 2017). Most of these chronobiological studies focus on the biological clock, known as the circadian rhythm. This is an autoregulatory system whose rhythm persists for approximately 24h without external time cues (Vitaterna, Takahashi, & Turek, 2001). An important feature of the circadian rhythm is that it shows temperature compensation, which the oscillator can maintain in a constant rhythm within a physiological range of temperatures, unlike the usual biochemical reaction (Bodenstein, Heiland, & Schuster, 2012). Another important feature is that the rhythm shows entrainment to pulses of stimulation such as light (Bodenstein et al., 2012). In this chapter, we focus on not only well-defined circadian rhythm and its related phenomenon, but also more broad diel responses observed in microbial world. Historically, since the fundamental energy source—sunlight—changes daily and it especially affects diel patterns of metabolic activities in photoautotrophic organisms (Bell-Pedersen et al., 2005), physiological responses of cultured model species such as eukaryotic microalgae and cyanobacteria have been studied (Nelson & Brand, 1979; Sournia, 1975). In parallel with this, diel observations of environmental microbial abundance and activity have also been made from an ecological perspective from the early time of microbial ecology (Vaulot, Marie, Olson, & Chisholm, 1995).

Recent developments and application of different omics techniques, and in particular environmental transcriptomics and metabolomics, together with co-occurrence network analysis is expanding our understanding of the existing interactions between different microorganisms in natural environments at different temporal scales (Faust & Raes, 2012). More importantly, these approaches provide a glimpse into how complex interaction networks and co-abundance patterns are influenced by the diel cycle constraints of cellular metabolism (and vice versa), leading to "fine-scale" pictures of environmental microbial diel activity at population level. These provide insights into a mechanistic understanding of how diel cycle-driven microbial physiological responses shape the community diel rhythm in aquatic ecosystems. In this chapter we have summarized recent understanding of microbial diel responses generated through complicated interactions in the environment.

2. Culture-based understanding of microbial diel patterns at each trophic level

Microbial diel dynamics were historically studied in cultured model species focusing mainly on their genetic and physiological responses (Nelson & Brand, 1979; Sournia, 1975). Thus, to gain a mechanistic understanding of how diel cycle-driven cellular metabolism might shape community diel dynamics in aquatic ecosystems, we have first summarized available experimental studies from individual species belonging to different trophic levels.

2.1 Photoautotrophic microorganisms

Diel periodicity in photoautotrophic microorganisms is a representative example of how biological patterns are shaped by direct interactions with the oscillating light energy. From the early stages of microbial diel cycle studies, periodically oscillating cell division, photosynthetic capacity and chlorophyll concentrations were reported in various species of microalgal cultures among divergent eukaryotic taxa such as diatoms, coccolithophores, cryptophytes, dinoflagellates and chlorophytes (Nelson & Brand, 1979). For example, an early comparable study reported observations of cell division in four diatom cultures during the light period, whereas another study tested eukaryotic microalgae such as coccolithophores (e.g., *Emiliania huxleyi*), dinoflagellates and chlorophytes showed the maximum division rate during the dark period (Nelson & Brand, 1979). In addition, intraspecific differences in division periodicity were also observed among eight clones of the diatom *Thalassiosira pseudonana* and six clones of the coccolithophore *Emiliania huxleyi* (Nelson & Brand, 1979).

In prokaryotes, cyanobacteria show diel periodicity in their metabolism and have been investigated as a model system of the biological clock. To date, a number of genetic and physiological studies have revealed diel synchronicity of their cellular processes such as substrate uptake (Mary et al., 2008), pigment ratio, optical properties (Bruyant et al., 2005; Claustre et al., 2002) and gene expression (Waldbauer, Rodrigue, Coleman, & Chisholm, 2012; Zinser et al., 2009). For example, in *Prochlorococcus*, DNA replication is known to occur in the afternoon followed by cell division at night (Vaulot et al., 1995).

Several diel responses in phototrophic organisms, such as cell division, nitrogen fixation and amino acid uptake in *Synechococcus*, are known to regulated by a circadian clock. This clock consists of at least one internal autonomous oscillator forming positive and negative feedback loops, with a cycle of approximately 24 h in many cases (Cohen & Golden, 2015; Reppert & Weaver, 2002; Young & Kay, 2001). Molecular components of circadian clock receive environmental input to synchronize with environmental time, then transmit temporal information to the rhythm-controlled output pathway. They thereby produce daily rhythmic behavior, enabling the cell to prepare for and take advantage of the daily environmental cycles (Bell-Pedersen et al., 2005).

The molecular mechanism of the microbial circadian clock was first discovered in an excellent genetic analysis of a circadian clock mutant of the cyanobacterial strain *Synechococcus elongatus* PCC 7942 (Kondo et al., 1994). Cyanobacterial circadian clocks are composed of three core oscillator genes, *kaiA*, *kaiB* and *kaiC* (Ishiura et al., 1998). Of these, a self-negative feedback control of *kaiC* gene expression is known to generate a circadian oscillation in cyanobacteria, whereas KaiA sustains diel oscillation by the enhancement of *kaiC* expression (Ishiura et al., 1998). Interestingly, *Prochlorococcus* strains have a truncated or completely absent *kaiA* gene in their streamlined genomes (Axmann et al., 2009). Their broken clocks, therefore, require daily light entrainment to function as a 24 h timer. This differs from a complete circadian clock that continues to function as an oscillator for multiple days under constant light conditions (Holtzendorff et al., 2008). Thus, photoautotrophic microorganisms form daily biological rhythms by a direct interaction with natural light.

In comparison with cyanobacteria, the molecular mechanism of circadian clocks in eukaryotic phototrophic microorganisms has been less documented. A biochemically well-studied example of a circadian clock in unicellular microalgae is *Lingulodinium polyedra* (a dinoflagellate; basionym *Gonyaulax polyedra*). Several biological functions of *L. polyedra* exhibit circadian rhythmicity, including motility, cell division, photosynthesis and bioluminescence (McMurry & Hastings, 1972; Njus, Van Gooch, & Hastings, 1981). These function can maintain about 24-h rhythmicity for several weeks under constant conditions (McMurry & Hastings, 1972; Njus et al., 1981). Interestingly, the biological rhythm of *L. polyedra* is known to be shortened by mammalian cell extracts (in a dose-dependent manner) by as much as 4h per day (Roenneberg, Nakamura, & Hastings, 1988). Creatine, which is involved in cellular energy metabolism in animal systems, was shown to be the limiting substance. A chemically similar endogenous limiting substance is also found in *L. polyedra*, and has been named "gonyaulin" (Roenneberg & Taylor, 1994). The effect of adding creatine is 100-fold greater than that of adding gonyaulin, suggesting that *L. polyedra* cells contain a certain amount of gonyaulin (Roenneberg & Taylor, 1994). Notably, a RNA-seq analysis has suggested that *L. polyedra* lacks detectable genome-wide daily variation in transcript levels (Roy et al., 2014). This indicates that the endogenous circadian timer of *L. polyedra* does not require rhythmic RNA. Therefore, *L. polyedra* may control its cellular time without transcriptional changes such as translational and post-translational control unlike well-known molecular biological clock systems such as the *kaiABC* in cyanobacteria.

The model organism *Chlamydomonas reinhardtii* (Chlorophyta) has been relatively well studied (Noordally & Millar, 2015). It possesses clock modulators such as XRN1 and CHLAMY1 that control the transcription of clock-relevant genes (Dathe, Prager, & Mittag, 2012; Matsuo et al., 2008), and phase angle and period of the circadian clock, respectively (Iliev et al., 2006), as well as photoreceptors such as cryptochromes and a phototropin (Mittag, Kiaulehn, & Johnson, 2005). Furthermore, the circadian clock of *Chlamydomonas* is composed of rhythms of chloroplast (roc) genes that share the conserved domains required for rhythmic regulation with *Arabidopsis thaliana* homologs (Brunner & Merrow, 2008; Matsuo et al., 2008; Matsuo, Iida, & Ishiura, 2012; Mittag et al., 2005). Further studies of the molecular mechanisms present in various microbial circadian clocks may provide more comprehensive knowledge about their diversity and universality.

2.2 Heterotrophic/mixotrophic microorganisms

While diel synchronicity of cellular processes in photoautotrophic microorganisms has been well investigated (see Section 2.1), there are only limited studies of diel periodicity in heterotrophic microorganisms. The periodicity of heterotrophic microorganisms has been observed in biomass and metabolic community activities (e.g., DOM, amino acids and CO_2), peaking during the day with some delay after phytoplankton activity

(Burney, Davis, Johnson, & Sieburth, 1982; Carlucci, Craven, & Henrichs, 1984; Fuhrman, Eppley, Hagström, & Azam, 1985; Johnson, Davis, & Sieburth, 1983). These early studies pointed out the relationship between heterotrophic periodicity and the activity of the primary producer; however, the mechanisms that exist to connect them have not been well characterized.

In heterotrophic prokaryotic microorganisms, a potential mechanism for the light-driven diel periodicity is a proton-pumping rhodopsin such as proteorhodopsin (PR). From the first discovery of PR in a genomic fragment of SAR86 proteobacteria in 2000 (Beja et al., 2000), the phototrophic strategy in heterotrophic microorganisms has received more attention. PR and relative microbial rhodopsin family contribute to a variety of cellular functions, such as ATP synthesis (Steindler, Schwalbach, Smith, Chan, & Giovannoni, 2011), substrate uptake (Gómez-Consarnau et al., 2016) and survival in low-nutrient conditions (Gómez-Consarnau et al., 2010) through light-driven ion pumping. These microbial rhodopsins are widely distributed within members of all three domains of life in surface aquatic ecosystems (Oesterhelt & Stoeckenius, 1973) and are also found in giant viruses (Yutin & Koonin, 2012). Environmental surveys have suggested that PRs are a major energy-transducing mechanism, comparable to photosynthesis, in the marine environment (Gómez-Consarnau et al., 2019; Olson, Yoshizawa, Boeuf, Iwasaki, & Delong, 2018). Although the relationships between community diel periodicity and PR activity are not documented, their energetical importance can be inferred by the community response to light conditions. Not all surface prokaryotes have such light-harvesting potential; however, a recent study suggested that several marine flavobacterial species without known light-harvesting mechanisms also exhibit metabolic change dependent on the visible light spectra (Hameed et al., 2020). This observation may suggest the importance of response to light conditions, even in such non-PR prokaryotes.

In eukaryotes, light-dependent grazing by mixotrophic microorganisms such as dinoflagellates have been well documented. Several dinoflagellates (*Flagilidium subglobosum*, *Gymnodinium gracilentum*, *Amphidinium poecilochroum*, *Dinophysis acuminata* and *Gyrodinium galatheanum*) are known to show high ingestion rates in light conditions compare with dark. In extreme case, some species ingest prey only in light conditions (Hansen & Nielsen, 1997; Jakobsen, Hansen, & Larsen, 2000; Kim et al., 2008; Li, Stoecker, & Coats, 2000). This light-dependent grazing is thought to be related to supplementation of major nutrients (N and P) for photosynthetic carbon assimilation (Li et al., 2000), or for efficiently digesting phytoplankton prey by utilizing its reactive oxygen species generated from the disruption of chloroplast structure and function (Strom, 2001). In mixotrophic species, growth and division cycles also show light-responding patterns similar to their phototrophic relatives. For example, the growth rate of *Fragilidium subglobosum* increases with light intensity ranging from 9 to 45 μ photons $m^{-2}s^{-1}$ (Hansen & Nielsen, 1997), while division rates in *Gymnodinium gracilentum* are higher than during the dark period (Jakobsen et al., 2000).

Such light-dependent grazing is also observed in the common marine ciliate *Mesodinium pulex* (Tarangkoon & Hansen, 2011). Recently, a potential mechanism between light and grazing relationship has been reported (Uzuka et al., 2019). When grazing unicellular photosynthetic prey, freshwater excavates (*Naegleria* spp.) and amebozoan (*Acanthamoeba* spp. and *Vannella* spp.) reduce their phagocytic predation but accelerate digestion under light conditions (Uzuka et al., 2019). A transcriptomic analysis suggests that this alternation contributes to reduce the phototoxicity during digestion of chlorophyll by minimizing the number of photosynthetic cells inside the predator cells (Uzuka et al., 2019). A biochemical strategy avoiding phototoxicity derived from chlorophylls has been reported in most of eukaryotic supergroups (Kashiyama et al., 2019). In these microeukaryotes, algal chlorophylls are rapidly converted to $13^2, 17^3$-cyclopheophorbide enols that are neither fluorescent nor photosensitive, resulting in the effective detoxification within the phagosomes (Kashiyama et al., 2012, 2013; Kinoshita, Kayama, Kashiyama, & Tamiaki, 2018). Such a catabolic process is phylogenetically ubiquitous among extant eukaryotes (Kashiyama et al., 2019). Thus, diel biological processes in heterotrophic/mixotrophic microorganisms may spread a diel rhythm to their habitat, by an indirect interaction with natural light via feeding on photosynthetic microorganisms.

2.3 Viruses

Viruses are the most abundant obligate parasites that depend on host metabolisms for their reproduction in marine environments (Suttle, 2005). Diel periodicity in these viruses has been mainly investigated in marine cyanoviruses infecting *Synechococcus* and *Prochlorococcus*, whereas little is known about diel cycling of viruses infecting heterotrophic bacteria by culture-based studies (see Section 3.3; their diel cycling by environmental survey).

Marine cyanoviruses are known to commonly possess multiple AMGs derived from host cyanobacteria (Ignacio-Espinoza & Sullivan, 2012; Mann et al., 2005; Thompson et al., 2011). For example, marine *Synechococcus* and *Prochlorococcus* viruses almost universally have the *psbA* gene encoding the D1 protein in the reaction center of Photosystem II (Clokie et al., 2006; Lindell, Jaffe, Johnson, Church, & Chisholm, 2005; Sullivan et al., 2006; Thompson et al., 2011). In addition, they often possess genes encoding the Photosystem II reaction center D2 (*psbD*) (Sullivan et al., 2006) and photosynthetic electron transport (*PTOX*, *petE* and *petF*) (Philosof, Battchikova, Aro, & Béjà, 2011; Sullivan et al., 2010).

These photosynthesis-related *AMG*s play central roles in ensuring energy for viral reproduction, by sustaining host photosynthetic activity after the shutoff of host metabolism during their infection processes (Clokie et al., 2006; Lindell et al., 2005; Mann, Cook, Millard, Bailey, & Clokie, 2003). Furthermore, to provide ATP and NADPH for viral genome replication, the pentose phosphate pathway is augmented while the Calvin cycle activity level decreases during viral infection (Thompson, Zeng, & Chisholm, 2016; Thompson et al., 2011). Thus, marine *Synechococcus* and *Prochlorococcus* viruses maintain host photosynthetic activity and redirect carbon flux from the Calvin cycle to the pentose phosphate pathway, using AMGs after the shutoff of host metabolisms for efficient viral reproduction (Clokie et al., 2006; Lindell et al., 2005; Sullivan et al., 2006; Thompson et al., 2011), with the exception of a few cyanoviruses showing a short latent period (Sullivan et al., 2010, 2006). As reflecting these genomic features, the gene expression of cyanoviruses reached peak levels linked to the host's photosynthetic activity, when the ATP level is the highest (Welkie et al., 2019), in the afternoon/dusk. This is followed by an increase in viral DNA in the free virions during the night (Aylward et al., 2017; Liu, Liu, Chen, Zhan, & Zeng, 2019; Yoshida et al., 2018). Furthermore, cyanoviruses are known to show light-dependent adsorption (Cséke & Farkas, 1979; Jia, Shan, Millard, Clokie, & Mann, 2010; Kao, Green, Stein, & Golden, 2005; Liu, Liu, et al., 2019) due to conformational changes in host receptors and/or viral tail fibers that can only be induced in the light (Jia et al., 2010). The other cyanoviruses with distinct genomic features (e.g., bloom-forming cyanobacterial viruses) show similar diel periodicity to that of marine cyanoviruses (Kimura et al., 2012; Morimoto et al., 2019; Morimoto, Kimura, Sako, & Yoshida, 2018). Thus, cyanoviruses form daily biological rhythms linked with light-utilizing host lifestyle.

3. Community diel cycles generate from complex biological interactions

We have seen that primary producers, including microalgae and cyanobacteria, exhibit diel cycles with a direct response to sunlight. Some heterotrophic microorganisms also have a potential light-harvesting or light-dependent grazing mechanism, and show—to some extent—self-sustaining diel periodicity. Even planktonic species inhabiting the photic zone of the world's oceans, and lacking a self-sustaining diel cycle regulation mechanism, indirectly exhibit community-level diel synchronization and dynamics (Ottesen et al., 2014; Tsai et al., 2012) by responses to the photosynthetic products released from primary producer cells. In this section, we focused on how the diel cycles originally produced in phototrophic microorganisms according to day–night light energy cycles are propagated to interacting heterotrophic community members, mostly through a food web comprising complex interactions, thereby generating a community diel cycle.

3.1 Tracking complex interactions within microbial communities

To better understand diel cycling in a natural microbial community, we should first review the complex microbial interactions taking place in the environment. Natural microbial communities are complex assemblages of different populations embedded in an even more complex network of ecological interactions. Fundamentally, these interactions can be either positive (e.g., cooperation or symbiosis), promoting microbial aggregation (e.g., cross-feeding, motility, quorum sensing and horizontal gene transfer); or negative (e.g., competition, parasitism, predation and infection), which eventually lead to the evolution of various defense and counter-defense strategies (e.g., allelopathy, biofilm formation, apoptosis and adaptive immunity) (Beliaev et al., 2014). Due to streamlined genomes and therefore limited transcriptional regulation capacity of most marine microorganisms (Swan et al., 2013) as well as prevailing nutrient limited environmental conditions in oceanic ecosystems, physiological and metabolic co-adaptation of interacting species on a diel-scale may be crucial to optimize their short-term performance and survival rates within the community.

Within a microbial community, for example, diverse organic matter (e.g., nutrients, enzymes, reactive oxygen species, toxins and high molecular weight compounds) are always released from each microorganism via biogenic activity and interactions (Beliaev et al., 2014; Buchan et al., 2014; Dziga, Maksylewicz, Maroszek, & Marek, 2018; Morris, Kirkegaard, Szul, Johnson, & Zinser, 2008; Seymour, Amin, Raina, & Stocker, 2017). Therefore, co-adaptation of different metabolisms can be advantageous for co-occurring species, by enabling them to uptake otherwise inaccessible energy and/or nutrients, or to respond in a timely manner to emerging stress. However, microbial interactions can switch from being cooperative to competitive because of density-dependent processes (Aharonovich & Sher, 2016), changes in cell physiology (healthy vs senescent) (Seyedsayamdost, Case, Kolter, & Clardy, 2011) or emergence of functional overlap (e.g., production and release of the compounds that can substitute for each other) among the co-occurring species (thus no need for cooperation) (Morris, Lenski, & Zinser, 2012), or a combination of these.

In general, positive interactions will be mirrored by: (i) the time-lagged correlations in transcript profiles of genes encoding basic cellular growth and division functions (e.g., ribosomal proteins, various polymerases, translation and transcription processing factors, as well as those involved in energy production and storage such as ATP synthases); or (ii) downregulation of specific genes associated with utilization of compounds provided by the co-occurring microbe (Beliaev et al., 2014). On the other hand, negative interactions can be characterized by co-expression of genes involved in stress response, DNA repair (e.g., exonucleases, DNA ligases and DNA photolyases) and various defense mechanisms (e.g., TA systems, R-M systems, Abi and CRISPR-Cas) in one species and genes involved in cell-to-cell signaling, motility, and allelopathic and proteolytic activities in the other (Aharonovich & Sher, 2016; Amin, Parker, & Armbrust, 2012; Doron et al., 2016). Thus, microorganisms are shackled by complex biotic interactions, and therefore elucidation of the specific transcriptional signatures reflecting these interactions is crucial for better understanding of the diel dynamics of microbial communities.

3.2 Diel cycling of microbial communities generated by direct interactions between phototrophic and heterotrophic microorganisms

In the photic zone of aquatic ecosystems, photoautotrophic microorganisms interact directly with natural light, and thereby transmit energy and carbon into microbial communities (molecular mechanisms are reviewed in Section 2.1) (Fig. 1A). Such diel responses start from the transcription of genes encoding parts of the photosynthesis apparatus in eukaryotic microalgae or cyanobacteria before sunrise, ensuring that cells are ready for energy generation as soon as light becomes available (Hellweger, Jabbur, Johnson, van Sebille, & Sasaki, 2020). This is followed by the upregulation of genes involved in energy storage (e.g., ATPases), growth (e.g., ribosomal and RNA polymerases) and DNA replication and repair during the daytime (Ottesen et al., 2013). Thus, numerous photoautotroph genes show diel expression patterns in their environments.

The products that are generated by these processes—such as reduced carbon (e.g., sugars, sugar alcohols and organic acids) and in diazotrophic cyanobacteria also nitrogen (e.g., ammonium, dissolved organic nitrogen) compounds—are partly released from algal or cyanobacterial cells via excretion, and then actively assimilated by the other heterotrophic/mixotrophic microorganisms (Seymour et al., 2017) (Fig. 1B). Such photosynthesis-driven pulse release of organic matter induces transcriptional shifts in heterotrophs by an increased upregulation of a number of both total and taxon-specific genes (e.g., ABC transporter) and metabolic pathways (e.g., TCA and methylcitrate cycles) involved in chemical transformation and uptake of organic matter (Aylward et al., 2015; Frischkorn, Haley, & Dyhrman, 2018; McCarren et al., 2010; Straub, Quillardet, Vergalli, de Marsac, & Humbert, 2011). Likewise, exudation of photosynthesis products has also been shown to promote expression of bacterial genes associated with motility and surface attachment (Azam & Malfatti, 2007; Beliaev et al., 2014; Geng & Belas, 2010) (Fig. 1B). These phytoplankton–bacteria interspecies interactions have been reported from various cultured phytoplankton–bacteria pairs such as *Synechococcus–Shewanella* (Beliaev et al., 2014), *Prochlorococcus*–SAR11 (Becker, Hogle, Rosendo, & Chisholm, 2019), *Synechococcus–Roseobacter* (Christie-Oleza, Sousoni, Lloyd, Armengaud, & Scanlan, 2017) and *Roseobacter*-diatom (*Thalassiosira pseudonana*) and dinoflagellate (*Alexandrium tamarense*). This suggests that specific phytoplankton–bacteria interactions may be common phenomena in marine environments as well and be not one-to-one but one-to-multiple species combination. For example, co-culture experiment of *Ruegeria pomeroyi* (*Roseobacter*) with *Alexandrium tamarense* or *Thalassiosira pseudonana* suggest *R. pomeroyi* show distinctive transcriptomic response to each phytoplankton species according to difference in exometabolites derived from phytoplankton (e.g., dinoflagellate released dimethysulfoniopropionate (DMSP), taurine, methylated

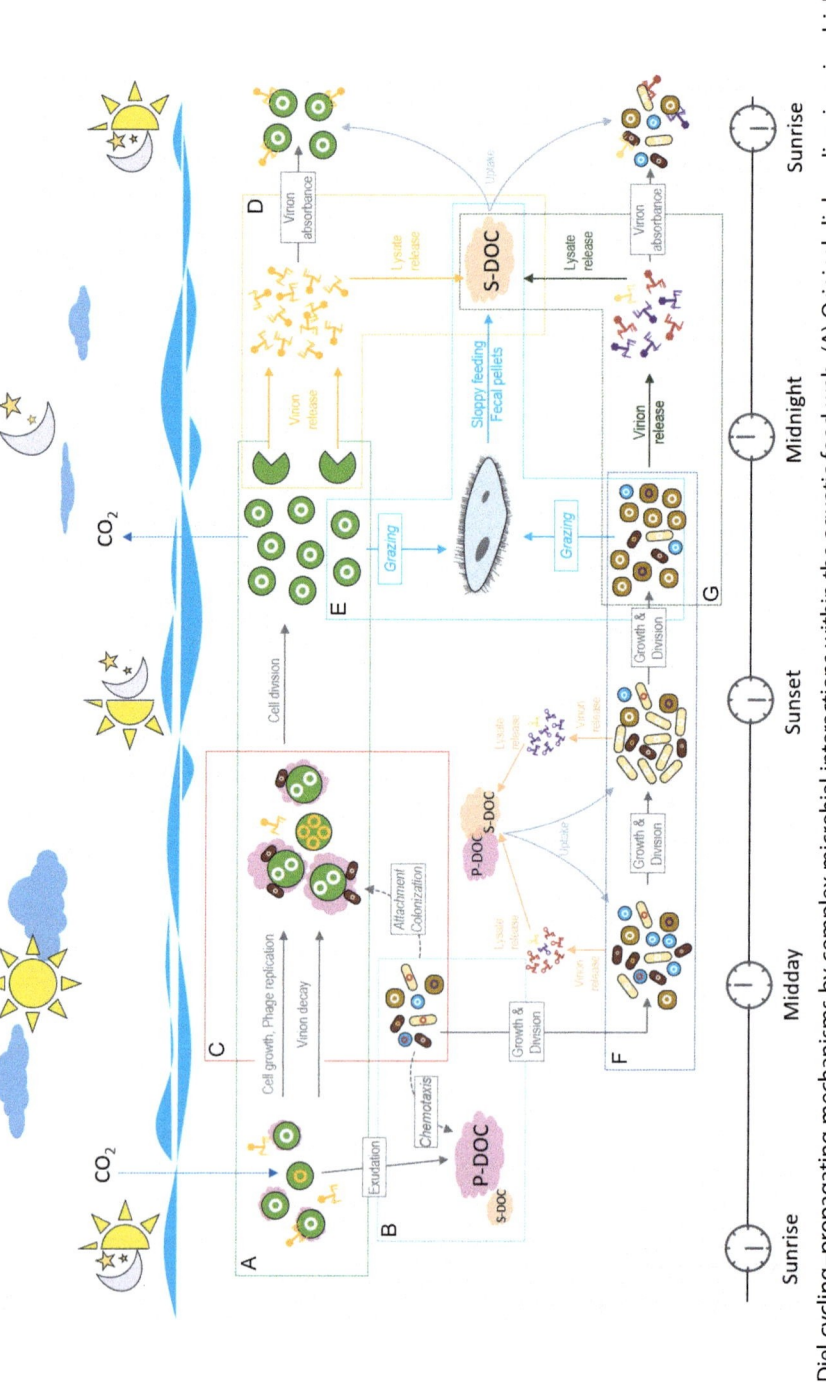

Fig. 1 Diel cycling–propagating mechanisms by complex microbial interactions within the aquatic food web. (A) Original diel cycling in microbial phototrophs in response to sunlight. (B) Photosynthesis-driven pulse release of organic matter transmits toward microbial heterotrophs. (C) Positive feedback from microbial heterotrophs may affect diel cycling in microbial phototrophs. (D) Release and diversity of organic matter from phytoplankton cells are affected by viral lysis. (E) Protistan grazing provides an additional source of organic matter. (F) Partitioning of niche space or differences in diel cycle

amines, and polyamines. Diatom mainly released dihydroxypropanesulfonate (DHPS), xylose, ectoine, and glycolate) (Landa, Burns, Roth, & Moran, 2017). These biotic interactions, based on nutrient availability, affect heterotrophic behavior and partly generate diel gene expression patterns in heterotrophs.

Heterotrophic microorganisms (e.g., *Alteromonas*) are also known to induce higher expression of genes exhibiting diel patterns in phototrophs (e.g., *Prochlorococcus*) when co-cultured with each other (Biller, Coe, Roggensack, & Chisholm, 2018), suggesting increased coordination of their metabolisms. This may be explained by a positive feedback mechanism driving mutual co-existence, which may occur owing to bacterial growth-mediated environmental changes at the cell surface interface of microalgae and cyanobacteria (e.g., variations in redox potential due to changes in pH and gas concentration), in turn enhancing utilization of the limiting trace metals (e.g., Fe, Mn and Mg) and increasing carbon and nitrogen fixation efficiency in primary producers (Basu, Gledhill, de Beer, Prabhu Matondkar, & Shaked, 2019) (Fig. 1C). In addition, researchers have observed increased expression of bacterial genes associated with the subsequent release of different ligands (e.g., siderophores and signaling molecules), co-enzymes (e.g., Acetyl-CoA and vitamins) and other growth-promoting compounds (Seyedsayamdost et al., 2011) that can stimulate DNA and protein metabolism in phototrophic organisms as a response to (and thus as an indication of) the emerging beneficial interactions (Aharonovich & Sher, 2016) (Fig. 1C). Hence, biotic interactions from heterotrophs toward phototrophs may also contribute to shape diel cycling in phototrophs, although further studies are needed to reveal the extent of the effect on them (Fig. 1C). Altogether, direct taxon-specific interaction via excretion or exudation (and probably vice versa) is one of the major mechanisms propagating and generating diel cycling in heterotrophs, with phototrophs as a starting point (Fig. 1A–C).

3.3 Diel cycling of microbial communities generated by indirect interactions

Direct interactions between phototrophs and heterotrophs are limited only to those species that can directly utilize photosynthetic products usually through the colonization of primary producers (Aylward et al., 2015; Frischkorn et al., 2018; McCarren et al., 2010; Seymour et al., 2017; Straub et al., 2011). In many cases, the diel cycling originating from a phototrophic microorganism is propagated relatively indirectly toward heterotrophic microorganisms as a result of complex biotic interactions within microbial loops (Deng, Cheung, & Liu, 2020; Fang et al., 2019; Hu, Connell, Mesrop, & Caron, 2018; Zhao et al., 2019). In this process, phototroph-derived organic matter plays an important role in transmitting and generating diel cycling in heterotrophs. For example, in *Prochlorococcus*—which is a dominant cyanobacterium in the marine environment—cell production and mortality rates are known to be tightly synchronized to the day/night cycle (Ribalet et al., 2015). Such diel cycling of organic matter contributes to shape community diel rhythms in aquatic ecosystems.

Viruses are major factors affecting such diel cycling of organic matter released from phototrophic microorganisms (Figs. 1D and 2). In Section 2.3, for example, we describes how cyanoviruses reproduce viral progeny linked with photosynthetic activity during the daytime, and then lyse host cyanobacterial cells at night, resulting in the generation of diel cycling of organic matter (Aylward et al., 2017; Kimura et al., 2012; Liu, Liu, et al., 2019; Morimoto et al., 2019; Welkie et al., 2019; Yoshida et al., 2018). Furthermore, cyanoviruses redirect host metabolism for efficient viral reproduction during the infection (Hurwitz & U'Ren, 2016; Puxty, Millard, Evans, & Scanlan, 2015; Zimmerman et al., 2019), suggesting that organic matter content can be affected by viral infection.

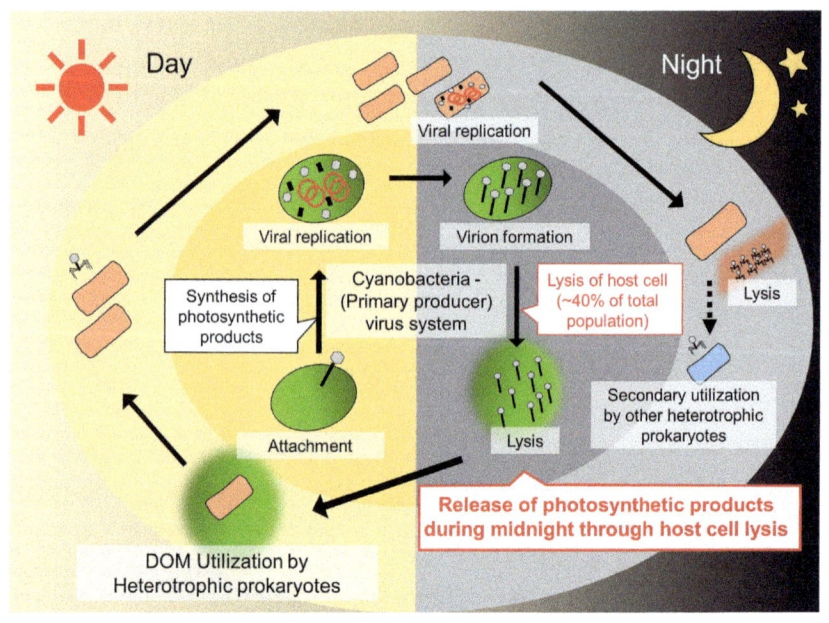

Fig. 2 Virus-induced organic matter release affecting diel cycling in a microbial community.

Such DOM, released via viral infection, indeed increases the diversity of DOM and induces the succession of heterotrophic microbial composition (Zhao et al., 2019) as well as altering gene expression in non-infected cells (Fang et al., 2019). Thus, viral-induced diel cycling of DOM and the resultant specific responses shape taxon-specific diel cycling in heterotrophic microorganisms in the environment (Figs. 1D and 2).

Another important factor affecting diel cycling in heterotrophic microorganisms is protistan grazing (Deng et al., 2020; Hu et al., 2018) (Fig. 1E). This process may reflect temporal adaptations to optimize resource availability. For example, it has been suggested that, during the daytime, the stoichiometric composition of the unicellular cyanobacterium *Crocosphaera* does not meet the nutrient demand of the protistan grazer; it is not until the cyanobacterium starts to fix atmospheric nitrogen during the dark period that it becomes preferentially grazed (Deng et al., 2020). Another adaptation to nighttime grazing activity may be associated with the reduced photosynthetic oxidative stress that predators may experience when ingesting phototrophic prey during the day (Uzuka et al., 2019). Such feeding during the night period provides additional DOM resources for the remaining photo- and heterotrophic community, leading to the induction of their response and generation of diel cycling (Fig. 1E). Indeed, Kelly et al. (2019) found increased diversity of genes associated with degradation of various carbohydrates at night. Altogether, diel cycling of phototrophic microorganisms is propagated toward heterotrophic microorganisms as a form of organic matter via viral lysis, or protistan grazing within the microbial community: a major mechanism generating diel rhythms for microorganisms that cannot utilize photosynthetic products as they are.

This section has shown that the primary diel cycle generated in phytoplankton according to the day–night cycle is transmitted to heterotrophic microbial organisms by direct or indirect microbial interactions. This microbial "interactions-driven" diel cycle generation not only seems to be specific to phototroph and heterotroph, but may also occur in heterotroph–heterotroph interactions (Fig. 1F). For example, different timings in expression maxima of genes associated with cellular activity were observed between co-occurring members of marine heterotrophic prokaryotes such as SAR11, SAR116 and SAR324 clades (Vislova, Sosa, Eppley, Romano, & DeLong, 2019). Although this may indicate simple temporal partitioning of their niche space due to competition avoidance, it may also be possible that this time lag was generated from the difference in the diel cycle transmitting pathway through microbial interactions in each taxon (e.g., differences in the number or species involved in the interactions). Consequently, this also suggests that bacteria exhibit higher diel variation in timing of the expression of different functional gene groups compared to phototrophic organisms. Recent studies revealed that transcriptional activity of individual populations of viruses is synchronized with their putative hosts (Aylward et al., 2017; Kolody et al., 2019; Martinez-Hernandez et al., 2020; Yoshida et al., 2018). Therefore, the viral lysis-mediated supply of organic matter multiply occurs in a day with different rhythmic pattern, and it may play a role in modulating the time lags of time partitioning in different heterotrophic taxa (Figs. 1G and 2). Because of the taxon-specific differences in nutrient requirements, the diel cycle of each heterotrophic taxon can depend on its specific supplier's diel lysis cycle with delay (Fig. 1D–E). Although dominant species of primary producer were shown to be as central determinants of overall community diel transcriptome dynamics (Aylward et al., 2015), considering the fundamental importance of microbial interaction for community structuring (Liu, Debeljak, Rembauville, Blain, & Obernosterer, 2019; Needham & Fuhrman, 2016), these heterotroph–heterotroph interaction also seems to be a modulating factor of diel community dynamics.

4. Conclusions and perspectives

In the photic zone of aquatic ecosystems, microorganisms with different metabolisms and their viruses form complex interactions and food webs. Within these interactions, phototrophic microorganisms such as microalgae and cyanobacteria interact directly with sunlight, and thereby generate circadian rhythms (Fig. 1A). The molecular mechanisms of circadian clock have been extensively studied in cyanobacteria and to a lesser degree in other primary producers. However, microbial primary producers transmit energy and carbon to the microbial community, in turn directly or indirectly propagate diel cycling toward the community. Photosynthesis-driven pulse release via excretion and exudation induces specific upregulation of genes and metabolic pathways, and thereby partly generates diel gene expression patterns in heterotrophic microorganisms (Fig. 1B). In addition, the diel cycling, originally generated in phototrophic microorganisms, also propagates toward heterotrophic microorganisms via more complex biotic interactions. Viral lysis-driven and protistan grazing-driven organic matter release during the night plays an important role in this process (Figs. 1D–E and 2). Cell death provides microorganisms that cannot directly utilize the photosynthetic products with diverse and additional resources of DOM, leading to the generation of diel responses (Fig. 1D–E). Likewise, it is possible that differences in the diel transmitting pathway through complex interactions among heterotrophs also generates higher variations and time lag diel responses in each taxon (Fig. 1F). Viruses infecting heterotrophic microorganisms and their mediated organic matter release may be important for modulating the time partitioning in the response of different heterotrophic taxa (Figs. 1G and 2). Together, sunlight and photosynthesis are important sources of diel cycling for aquatic microorganisms, as well as for supplying energy and carbon.

These diel oscillations in microbial activity have been observed in various photic aquatic environments such as coastal water, open ocean and coral reefs, and also in freshwater (Aylward et al., 2017; Kelly et al., 2019; Larkin et al., 2020; Ottesen et al., 2014). The next areas of research in this field will be to visualize the diffusion of this diel rhythm into the aphotic zone, and then to reveal the spatiotemporal distribution and patterns of diel oscillating phenomena. A recent study tackled this problem, and observed the decline in transcript oscillations in a mesopelagic water community compared with those of the water surface (Vislova et al., 2019). However, if the connections between microbial communities on the surface and in deep layers are considered from the perspective of carbon sinking, the diel oscillation of a surface community can influence the rhythm of a deep layer community under some specific environmental conditions. For example, during a phytoplankton bloom, active photosynthetic productivity and release of organic matter may make a greater contribution to the rhythm at the deeper layer than is made under non-bloom conditions. Further studies are needed to clarify the spatial and temporal change in rhythmic diel patterns on the aquatic microbial community.

In addition, these microbial organisms interact not only with other members of the microbial community, but also potentially with higher organisms such as zooplankton (e.g., copepoda) in the environment. It is well known that these higher aquatic organisms also show rhythmic diel patterns. For example, copepods show a diel feeding pattern with vertical migration (Atkinson, Ward, Williams, & Poulet, 1992; Dagg, Frost, & Walser, 1989; Durbin, Durbin, & Wlodarczyk, 1990). The diel activity patterns of these zooplankton are driven by light intensity changes and related to factors such as UV radiation, predation risk and availability of their microbial food (Ringelberg, 1995; Stearns & Forward, 1984; Williamson, Fischer, Bollens, Overholt, & Breckenridgec, 2011). These diel patterns in higher organisms can be directly or indirectly influenced by microbial diel cycles and vice versa. As is the case with plants in terrestrial environments, microbial phytoplankton take on the central responsibility in entire marine and freshwater ecosystems. This implies the larger potential importance of microbial organisms and also of their diel rhythms at least in aquatic ecosystems.

Finally, microbial interactions in marine environments are influenced not only by the day–night cycle but also by other environmental cycles, which are mostly linked to the periodical movements of the sun or the moon. Future work to reveal how season-related (e.g., monthly changes in tidal cycles and moonlight conditions, and seasonal changes in day length) changes in the day–night cycle shape microbial community diel responses is important to fully comprehend community dynamics and biogeochemical cycling in aquatic environments.

Acknowledgments

D.M., S.Š., K.T. and T.Y. all contributed to manuscript preparation, discussion, and revision. We thank Edanz Group (www.edanzediting.com/ac) for language editing of a draft version of this manuscript.

References

Aharonovich, D., & Sher, D. (2016). Transcriptional response of Prochlorococcus to co-culture with a marine Alteromonas: Differences between strains and the involvement of putative infochemicals. *The ISME Journal, 10*, 2892–2906. https://doi.org/10.1038/ismej.2016.70.

Alonso-Sáez, L., Díaz-Pérez, L., & Morán, X. A. G. (2015). The hidden seasonality of the rare biosphere in coastal marine bacterioplankton. *Environmental Microbiology, 17*, 3766–3780. https://doi.org/10.1111/1462-2920.12801.

Amin, S. A., Parker, M. S., & Armbrust, E. V. (2012). Interactions between diatoms and bacteria. *Microbiology and Molecular Biology Reviews, 76*, 667–684. https://doi.org/10.1128/mmbr.00007-12.

Atkinson, A., Ward, P., Williams, R., & Poulet, S. A. (1992). Feeding rates and diel vertical migration of copepods near South Georgia: Comparison of shelf and oceanic sites. *Marine Biology, 114*, 49–56. https://doi.org/10.1007/BF00350855.

Axmann, I. M., Dühring, U., Seeliger, L., Arnold, A., Vanselow, J. T., Kramer, A., et al. (2009). Biochemical evidence for a timing mechanism in Prochlorococcus. *Journal of Bacteriology, 191*, 5342–5347. https://doi.org/10.1128/JB.00419-09.

Aylward, F. O., Boeuf, D., Mende, D. R., Wood-Charlson, E. M., Vislova, A., Eppley, J. M., et al. (2017). Diel cycling and long-term persistence of viruses in the ocean's euphotic zone. *Proceedings of the National Academy of Sciences of the United States of America, 114*, 11446–11451. https://doi.org/10.1073/pnas.1714821114.

Aylward, F. O., Eppley, J. M., Smith, J. M., Chavez, F. P., Scholin, C. A., & DeLong, E. F. (2015). Microbial community transcriptional networks are conserved in three domains at ocean basin scales. *Proceedings of the National Academy of Sciences of the United States of America, 112*, 5443–5448. https://doi.org/10.1073/pnas.1502883112.

Azam, F., Fenchel, T., Field, J. G., Gray, J. S., Meyer-Reil, L. A., & Thingstad, F. (1983). The ecological role of water-column microbes in the sea. *Marine Ecology Progress Series*, *10*, 257–263. https://doi.org/10.3354/meps010257.

Azam, F., & Malfatti, F. (2007). Microbial structuring of marine ecosystems. *Nature Reviews. Microbiology*, *5*, 782–791. https://doi.org/10.1038/nrmicro1747.

Bar-On, Y. M., Phillips, R., & Milo, R. (2018). The biomass distribution on earth. *Proceedings of the National Academy of Sciences of the United States of America*, *115*, 6506–6511. https://doi.org/10.1073/pnas.1711842115.

Basu, S., Gledhill, M., de Beer, D., Prabhu Matondkar, S. G., & Shaked, Y. (2019). Colonies of marine cyanobacteria Trichodesmium interact with associated bacteria to acquire iron from dust. *Communications Biology*, *2*, 1–8. https://doi.org/10.1038/s42003-019-0534-z.

Becker, J. W., Hogle, S. L., Rosendo, K., & Chisholm, S. W. (2019). Co-culture and biogeography of Prochlorococcus and SAR11. *The ISME Journal*, *13*, 1506–1519. https://doi.org/10.1038/s41396-019-0365-4.

Beja, O., Aravind, L., Koonin, E. V., Suzuki, M. T., Hadd, A., Nguyen, L. P., et al. (2000). Bacterial rhodopsin: Evidence for a new type of phototrophy in the sea. *Science (80-)*, *289*, 1902–1906. https://doi.org/10.1126/science.289.5486.1902.

Beliaev, A. S., Romine, M. F., Serres, M., Bernstein, H. C., Linggi, B. E., Markillie, L. M., et al. (2014). Inference of interactions in cyanobacterial-heterotrophic co-cultures via transcriptome sequencing. *The ISME Journal*, *8*, 2243–2255. https://doi.org/10.1038/ismej.2014.69.

Bell-Pedersen, D., Cassone, M. V., Earnest, J. D., Golden, S. S., Hardin, E. P., Thomas, L. T., et al. (2005). Circadian rhythms from multiple oscillators: Lessons from diverse organisms. *Nature Reviews. Drug Discovery*, *6*, 544–556. https://doi.org/10.1038/nrd1633.

Bhadra, U., Thakkar, N., Das, P., & Pal Bhadra, M. (2017). Evolution of circadian rhythms: From bacteria to human. *Sleep Medicine*, *35*, 49–61. https://doi.org/10.1016/j.sleep.2017.04.008.

Biller, S. J., Coe, A., Roggensack, S. E., & Chisholm, S. W. (2018). Heterotroph interactions alter Prochlorococcus Transcriptome dynamics during extended periods of darkness. *mSystems*, *3*, 1–18. https://doi.org/10.1128/msystems.00040-18.

Bodenstein, C., Heiland, I., & Schuster, S. (2012). Temperature compensation and entrainment in circadian rhythms. *Physical Biology*, *9*, 36011. https://doi.org/10.1088/1478-3975/9/3/036011.

Breitbart, M., Bonnain, C., Malki, K., & Sawaya, N. A. (2018). Phage puppet masters of the marine microbial realm. *Nature Microbiology*, *3*, 754–766. https://doi.org/10.1038/s41564-018-0166-y.

Breitbart, M., Thompson, L., Suttle, C., & Sullivan, M. (2007). Exploring the vast diversity of marine viruses. *Oceanography*, *20*, 135–139. https://doi.org/10.5670/oceanog.2007.58.

Brunner, M., & Merrow, M. (2008). The green yeast uses its plant-like clock to regulate its animal-like tail. *Genes & Development*, *22*, 825–831. https://doi.org/10.1101/gad.1664508.

Bruyant, F., Babin, M., Genty, B., Prasil, O., Behrenfeld, M. J., Claustre, H., et al. (2005). Diel variations in the photosynthetic parameters of Prochlorococcus strain PCC 9511: Combined effects of light and cell cycle. *Limnology and Oceanography*, *50*, 850–863. https://doi.org/10.4319/lo.2005.50.3.0850.

Buchan, A., LeCleir, G. R., Gulvik, C. A., & González, J. M. (2014). Master recyclers: Features and functions of bacteria associated with phytoplankton blooms. *Nature Reviews. Microbiology*, *12*, 686–698. https://doi.org/10.1038/nrmicro3326.

Bunse, C., & Pinhassi, J. (2017). Marine Bacterioplankton seasonal succession dynamics. *Trends in Microbiology*, *25*, 494–505. https://doi.org/10.1016/j.tim.2016.12.013.

Burney, C. M., Davis, P. G., Johnson, K. M., & Sieburth, J. M. N. (1982). Diel relationships of microbial trophic groups and in situ dissolved carbohydrate dynamics in the caribbean sea. *Marine Biology*, *67*, 311–322. https://doi.org/10.1007/BF00397726.

Carlucci, A. F., Craven, D. B., & Henrichs, S. M. (1984). Diel production and microheterotrophic utilization of dissolved free amino acids in waters off southern California. *Applied and Environmental Microbiology, 48*, 165–170. https://doi.org/10.1128/aem.48.1.165-170.1984.

Christie-Oleza, J. A., Sousoni, D., Lloyd, M., Armengaud, J., & Scanlan, D. J. (2017). Nutrient recycling facilitates long-term stability of marine microbial phototroph-heterotroph interactions. *Nature Microbiology, 2*, 17100. https://doi.org/10.1038/nmicrobiol.2017.100.

Claustre, H., Bricaud, A., Babin, M., Bruyant, F., Guillou, L., Le Gall, F., et al. (2002). Diel variations in Prochlorococcus optical properties. *Limnology and Oceanography, 47*, 1637–1647. https://doi.org/10.4319/lo.2002.47.6.1637.

Clokie, M. R. J., Shan, J., Bailey, S., Jia, Y., Krisch, H. M., West, S., et al. (2006). Transcription of a "photosynthetic" T4-type phage during infection of a marine cyanobacterium. *Environmental Microbiology, 8*, 827–835. https://doi.org/10.1111/j.1462-2920.2005.00969.x.

Cohen, S. E., & Golden, S. S. (2015). Circadian rhythms in cyanobacteria. *Microbiology and Molecular Biology Reviews, 79*, 373–385. https://doi.org/10.1128/mmbr.00036-15.

Collado-Fabbri, S., Vaulot, D., & Ulloa, O. (2011). Structure and seasonal dynamics of the eukaryotic picophytoplankton community in a wind-driven coastal upwelling ecosystem. *Limnology and Oceanography, 56*, 2334–2346. https://doi.org/10.4319/lo.2011.56.6.2334.

Cram, J. A., Chow, C. E. T., Sachdeva, R., Needham, D. M., Parada, A. E., Steele, J. A., et al. (2015). Seasonal and interannual variability of the marine bacterioplankton community throughout the water column over ten years. *The ISME Journal, 9*, 563–580. https://doi.org/10.1038/ismej.2014.153.

Cséke, C. S., & Farkas, G. L. (1979). Effect of light on the attachment of cyanophage AS-1 to Anacystis nidulans. *Journal of Bacteriology, 137*, 667 LP–669.

Dagg, M. J., Frost, B. W., & Walser, W. E. (1989). Copepod diel migration, feeding, and the vertical flux of pheopigments. *Limnology and Oceanography, 34*, 1062–1071. https://doi.org/10.4319/lo.1989.34.6.1062.

Dathe, H., Prager, K., & Mittag, M. (2012). Novel interaction of two clock-relevant RNA-binding proteins C3 and XRN1 in Chlamydomonas reinhardtii. *FEBS Letters, 586*, 3969–3973. https://doi.org/10.1016/j.febslet.2012.09.046.

Deng, L., Cheung, S., & Liu, H. (2020). Protistal grazers increase grazing on unicellular cyanobacteria diazotroph at night. *Frontiers in Marine Science, 7*, 1–10. https://doi.org/10.3389/fmars.2020.00135.

Doron, S., Fedida, A., Hernández-Prieto, M. A., Sabehi, G., Karunker, I., Stazic, D., et al. (2016). Transcriptome dynamics of a broad host-range cyanophage and its hosts. *The ISME Journal, 10*, 1437–1455. https://doi.org/10.1007/s13398-014-0173-7.2.

Ducklow, H. (2000). Bacterioplankton production and biomass in the oceans. In *Microbial ecology of the oceans* (1st ed., pp. 85–120). Wiley-Liss, Inc.

Durbin, A. G., Durbin, E. G., & Wlodarczyk, E. (1990). Diel feeding behavior in the marine copepod Acartia tonsa in relation to food availability. *Marine Ecology Progress Series, 68*, 23–45. https://doi.org/10.3354/meps068023.

Dziga, D., Maksylewicz, A., Maroszek, M., & Marek, S. (2018). Combined treatment of toxic cyanobacteria Microcystis aeruginosa with hydrogen peroxide and microcystin biodegradation agents results in quick toxin elimination. *Acta Biochimica Polonica, 65*, 133–140. https://doi.org/10.18388/abp.2017_2538.

Fang, X., Liu, Y., Zhao, Y., Chen, Y., Liu, R., Qin, Q. L., et al. (2019). Transcriptomic responses of the marine cyanobacterium Prochlorococcus to viral lysis products. *Environmental Microbiology, 21*, 2015–2028. https://doi.org/10.1111/1462-2920.14513.

Faust, K., Lahti, L., Gonze, D., de Vos, W. M., & Raes, J. (2015). Metagenomics meets time series analysis: Unraveling microbial community dynamics. *Current Opinion in Microbiology*, *25*, 56–66. https://doi.org/10.1016/j.mib.2015.04.004.

Faust, K., & Raes, J. (2012). Microbial interactions: From networks to models. *Nature Reviews. Microbiology*, *10*, 538–550. https://doi.org/10.1038/nrmicro2832.

Field, C. B., Behrenfeld, M. J., Randerson, J. T., & Falkowski, P. (1998). Primary production of the biosphere: Integrating terrestrial and oceanic components. *Science (80-)*, *281*, 237–240. https://doi.org/10.1126/science.281.5374.237.

Frischkorn, K. R., Haley, S. T., & Dyhrman, S. T. (2018). Coordinated gene expression between Trichodesmium and its microbiome over day–night cycles in the North Pacific subtropical gyre. *The ISME Journal*, *12*, 997–1007. https://doi.org/10.1038/s41396-017-0041-5.

Fuhrman, J. A., Cram, J. A., & Needham, D. M. (2015). Marine microbial community dynamics and their ecological interpretation. *Nature Reviews. Microbiology*, *13*, 133–146. https://doi.org/10.1038/nrmicro3417.

Fuhrman, J. A., Eppley, R. W., Hagström, Å., & Azam, F. (1985). Diel variations in bacterioplankton, phytoplankton, and related parameters in the Southern California bight. *Marine Ecology Progress Series*, *27*, 9–20.

Fuhrman, J. A., Hewson, I., Schwalbach, M. S., Steele, J. A., Brown, M. V., & Naeem, S. (2006). Annually reoccurring bacterial communities are predictable from ocean conditions. *Proceedings of the National Academy of Sciences of the United States of America*, *103*, 13104–13109. https://doi.org/10.1073/pnas.0602399103.

Geng, H., & Belas, R. (2010). Molecular mechanisms underlying roseobacter-phytoplankton symbioses. *Current Opinion in Biotechnology*, *21*, 332–338. https://doi.org/10.1016/j.copbio.2010.03.013.

Gómez-Consarnau, L., Akram, N., Lindell, K., Pedersen, A., Neutze, R., Milton, D. L., et al. (2010). Proteorhodopsin phototrophy promotes survival of marine bacteria during starvation. *PLoS Biology*, *8*, 2–11. https://doi.org/10.1371/journal.pbio.1000358.

Gómez-Consarnau, L., González, J. M., Riedel, T., Jaenicke, S., Wagner-Döbler, I., Sañudo-Wilhelmy, S. A., et al. (2016). Proteorhodopsin light-enhanced growth linked to vitamin-B 1 acquisition in marine Flavobacteria. *The ISME Journal*, *10*, 1102–1112. https://doi.org/10.1038/ismej.2015.196.

Gómez-Consarnau, L., Raven, J. A., Levine, N. M., Cutter, L. S., Wang, D., Seegers, B., et al. (2019). Microbial rhodopsins are major contributors to the solar energy captured in the sea. *Science Advances*, *5*, 1–7. https://doi.org/10.1126/sciadv.aaw8855.

Hameed, A., Lai, W. A., Shahina, M., Stothard, P., Young, L. S., Lin, S. Y., et al. (2020). Differential visible spectral influence on carbon metabolism in heterotrophic marine flavobacteria. *FEMS Microbiology Ecology*, *96*, 1–11. https://doi.org/10.1093/femsec/fiaa011.

Hansen, P. J., & Nielsen, T. G. (1997). Mixotrophic feeding of Fragilidium subglobosum (dinophyceae) on three species of Ceratium: Effects of prey concentration, prey species and light intensity. *Marine Ecology Progress Series*, *147*, 187–196. https://doi.org/10.3354/meps147187.

Hellweger, F. L., Jabbur, M. L., Johnson, C. H., van Sebille, E., & Sasaki, H. (2020). Circadian clock helps cyanobacteria manage energy in coastal and high latitude ocean. *The ISME Journal*, *14*, 560–568. https://doi.org/10.1038/s41396-019-0547-0.

Holtzendorff, J., Partensky, F., Mella, D., Lennon, J. F., Hess, W. R., & Garczarek, L. (2008). Genome streamlining results in loss of robustness of the circadian clock in the marine cyanobacterium Prochlorococcus marinus PCC 9511. *Journal of Biological Rhythms*, *23*, 187–199. https://doi.org/10.1177/0748730408316040.

Hu, S. K., Connell, P. E., Mesrop, L. Y., & Caron, D. A. (2018). A hard day's night: Diel shifts in microbial eukaryotic activity in the North Pacific subtropical gyre. *Frontiers in Marine Science*, 5, 1–17. https://doi.org/10.3389/fmars.2018.00351.

Hurwitz, B. L., & U'Ren, J. M. (2016). Viral metabolic reprogramming in marine ecosystems. *Current Opinion in Microbiology*, 31, 161–168. https://doi.org/10.1016/j.mib.2016.04.002.

Ignacio-Espinoza, J. C., Ahlgren, N. A., & Fuhrman, J. A. (2020). Long-term stability and red queen-like strain dynamics in marine viruses. *Nature Microbiology*, 5, 265–271. https://doi.org/10.1038/s41564-019-0628-x.

Ignacio-Espinoza, J. C., & Sullivan, M. B. (2012). Phylogenomics of T4 cyanophages: Lateral gene transfer in the "core" and origins of host genes. *Environmental Microbiology*, 14, 2113–2126. https://doi.org/10.1111/j.1462-2920.2012.02704.x.

Iliev, D., Voytsekh, O., Schmidt, E. M., Fiedler, M., Nykytenko, A., & Mittag, M. (2006). A heteromeric RNA-binding protein is involved in maintaining acrophase and period of the circadian clock. *Plant Physiology*, 142, 797–806. https://doi.org/10.1104/pp.106.085944.

Ishiura, M., Kutsuna, S., Aoki, S., Iwasaki, H., Andersson, C. R., Tanabe, A., et al. (1998). Expression of a gene cluster kaiABC as a circadian feedback process in cyanobacteria. *Science (80-)*, 281, 1519–1523. https://doi.org/10.1126/science.281.5382.1519.

Jakobsen, H. H., Hansen, P. J., & Larsen, J. (2000). Growth and grazing responses of two chloroplast-retaining dinoflagellates: Effect of irradiance and prey species. *Marine Ecology Progress Series*, 201, 121–128. https://doi.org/10.3354/meps201121.

Jia, Y., Shan, J., Millard, A., Clokie, M. R. J., & Mann, N. H. (2010). Light-dependent adsorption of photosynthetic cyanophages to Synechococcus sp. WH7803. *FEMS Microbiology Letters*, 310, 120–126. https://doi.org/10.1111/j.1574-6968.2010.02054.x.

Johnson, K. M., Davis, P. G., & Sieburth, J. M. N. (1983). Diel variation of TCO z in the upper layer of oceanic waters reflects microbial composition, variation and possibly methane cycling. *Marine Biology*, 77, 1–10. https://doi.org/10.1038/164914a0.

Kao, C. C., Green, S., Stein, B., & Golden, S. S. (2005). Diel infection of a Cyanobacterium by a contractile bacteriophage. *Applied and Environmental Microbiology*, 71, 4276 LP–4279. https://doi.org/10.1128/AEM.71.8.4276-4279.2005.

Kashiyama, Y., Yokoyama, A., Kinoshita, Y., Shoji, S., Miyashiya, H., Shiratori, T., et al. (2012). Ubiquity and quantitative significance of detoxification catabolism of chlorophyll associated with protistan herbivory. *Proceedings of the National Academy of Sciences of the United States of America*, 109, 17328–17335. https://doi.org/10.1073/pnas.1207347109.

Kashiyama, Y., Yokoyama, A., Shiratori, T., Hess, S., Not, F., Bachy, C., et al. (2019). Taming chlorophylls by early eukaryotes underpinned algal interactions and the diversification of the eukaryotes on the oxygenated earth. *The ISME Journal*, 13, 1899–1910. https://doi.org/10.1038/s41396-019-0377-0.

Kashiyama, Y., Yokoyama, A., Shiratori, T., Inouye, I., Kinoshita, Y., Mizoguchi, T., et al. (2013). 132,173-cyclopheophorbide b enol as a catabolite of chlorophyll b in phycophagy by protists. *FEBS Letters*, 587, 2578–2583. https://doi.org/10.1016/j.febslet.2013.06.036.

Kelly, L. W., Nelson, C. E., Haas, A. F., Naliboff, D. S., Calhoun, S., Carlson, C. A., et al. (2019). Diel population and functional synchrony of microbial communities on coral reefs. *Nature Communications*, 10, 1691. https://doi.org/10.1038/s41467-019-09419-z.

Kim, S., Kang, Y. G., Kim, H. S., Yih, W., Coats, D. W., & Park, M. G. (2008). Growth and grazing responses of the mixotrophic dinoflagellate Dinophysis acuminata as functions of light intensity and prey concentration. *Aquatic Microbial Ecology*, 51, 301–310. https://doi.org/10.3354/ame01203.

Kimura, S., Yoshida, T., Hosoda, N., Honda, T., Kuno, S., Kamiji, R., et al. (2012). Diurnal infection patterns and impact of Microcystis cyanophages in a Japanese pond. *Applied and Environmental Microbiology*, 78, 5805–5811. https://doi.org/10.1128/AEM.00571-12.

Kinoshita, Y., Kayama, M., Kashiyama, Y., & Tamiaki, H. (2018). In vivo and in vitro preparation of divinyl-132,173-cyclopheophorbide-a enol. *Bioorganic & Medicinal Chemistry Letters, 28*, 1090–1092. https://doi.org/10.1016/j.bmcl.2018.02.015.

Kolody, B. C., McCrow, J. P., Allen, L. Z., Aylward, F. O., Fontanez, K. M., Moustafa, A., et al. (2019). Diel transcriptional response of a California current plankton microbiome to light, low iron, and enduring viral infection. *The ISME Journal, 13*, 2817–2833. https://doi.org/10.1038/s41396-019-0472-2.

Kondo, T., Tsinoremas, N. F., Golden, S. S., Johnson, C. H., Kutsuna, S., & Ishiura, M. (1994). Circadian clock mutants of cyanobacteria. *Science (80-.), 266*, 1233–1236. https://doi.org/10.1126/science.7973706.

Landa, M., Burns, A. S., Roth, S. J., & Moran, M. A. (2017). Bacterial transcriptome remodeling during sequential co-culture with a marine dinoflagellate and diatom. *The ISME Journal, 11*, 2677–2690. https://doi.org/10.1038/ismej.2017.117.

Larkin, A. A., Garcia, C. A., Ingoglia, K. A., Garcia, N. S., Baer, S. E., Twining, B. S., et al. (2020). Subtle biogeochemical regimes in the Indian Ocean revealed by spatial and diel frequency of Prochlorococcus haplotypes. *Limnology and Oceanography, 65*, S220–S232. https://doi.org/10.1002/lno.11251.

Li, A., Stoecker, D. K., & Coats, D. W. (2000). Mixotrophy in Gyrodinium galatheanum (Dinophyceae): Grazing responses to light intensity and inorganic nutrients. *Journal of Phycology, 36*, 33–45. https://doi.org/10.1046/j.1529-8817.2000.98076.x.

Lindell, D., Jaffe, J. D., Johnson, Z. I., Church, G. M., & Chisholm, S. W. (2005). Photosynthesis genes in marine viruses yield proteins during host infection. *Nature, 438*, 86–89. https://doi.org/10.1038/nature04111.

Liu, Y., Debeljak, P., Rembauville, M., Blain, S., & Obernosterer, I. (2019). Diatoms shape the biogeography of heterotrophic prokaryotes in early spring in the Southern Ocean. *Environmental Microbiology, 21*, 1452–1465. https://doi.org/10.1111/1462-2920.14579.

Liu, R., Liu, Y., Chen, Y., Zhan, Y., & Zeng, Q. (2019). Cyanobacterial viruses exhibit diurnal rhythms during infection. *Proceedings of the National Academy of Sciences of the United States of America, 116*, 14077–14082. https://doi.org/10.1073/pnas.1819689116.

Mann, N. H., Clokie, M. R. J., Millard, A., Cook, A., Wilson, W. H., Wheatley, P. J., et al. (2005). The genome of S-PM2, a "photosynthetic" T4-type bacteriophage that infects marine Synechococcus strains. *Journal of Bacteriology, 187*, 3188–3200. https://doi.org/10.1128/JB.187.9.3188-3200.2005.

Mann, N. H., Cook, A., Millard, A., Bailey, S., & Clokie, M. (2003). Bacterial photosynthesis genes in a virus. *Nature, 424*, 741–742. https://doi.org/10.1038/424741a.

Marquardt, M., Vader, A., Stübner, E. I., Reigstad, M., & Gabrielsen, T. M. (2016). Strong seasonality of marine microbial eukaryotes in a high-Arctic. *Applied and Environmental Microbiology, 82*, 1868–1880. https://doi.org/10.1128/AEM.03208-15.Editor.

Martinez-Hernandez, F., Luo, E., Tominaga, K., Ogata, H., Yoshida, T., DeLong, E. F., et al. (2020). Diel cycling of the cosmopolitan abundant Pelagibacter virus 37-F6: One of the most abundant viruses on earth. *Environmental Microbiology Reports, 12*, 214–219. https://doi.org/10.1111/1758-2229.12825.

Mary, I., Garczarek, L., Tarran, G. A., Kolowrat, C., Terry, M. J., Scanlan, D. J., et al. (2008). Diel rhythmicity in amino acid uptake by Prochlorococcus. *Environmental Microbiology, 10*, 2124–2131. https://doi.org/10.1111/j.1462-2920.2008.01633.x.

Matsuo, T., Iida, T., & Ishiura, M. (2012). N-terminal acetyltransferase 3 gene is essential for robust circadian rhythm of bioluminescence reporter in Chlamydomonas reinhardtii. *Biochemical and Biophysical Research Communications, 418*, 342–346. https://doi.org/10.1016/j.bbrc.2012.01.023.

Matsuo, T., Okamoto, K., Onai, K., Niwa, Y., Shimogawara, K., & Ishiura, M. (2008). A systematic forward genetic analysis identified components of the Chlamydomonas circadian system. *Genes & Development, 22*, 918–930. https://doi.org/10.1101/gad.1650408.

McCarren, J., Becker, J. W., Repeta, D. J., Shi, Y., Young, C. R., Malmstrom, R. R., et al. (2010). Microbial community transcriptomes reveal microbes and metabolic pathways associated with dissolved organic matter turnover in the sea. *Proceedings of the National Academy of Sciences of the United States of America, 107,* 16420–16427. https://doi.org/10.1073/pnas.1010732107.

McMurry, L., & Hastings, J. W. (1972). No desynchronization among four circadian rhythms in the unicellular alga, Gonyaulax polyedra. *Science (80-), 175,* 1137 LP–1139. https://doi.org/10.1126/science.175.4026.1137.

Mittag, M., Kiaulehn, S., & Johnson, C. H. (2005). The circadian clock in Chlamydomonas reinhardtii. What is it for? What is it similar to? *Plant Physiology, 137,* 399–409. https://doi.org/10.1104/pp.104.052415.

Morimoto, D., Kimura, S., Sako, Y., & Yoshida, T. (2018). Transcriptome analysis of a bloom-forming cyanobacterium Microcystis aeruginosa during Ma-LMM01 phage infection. *Frontiers in Microbiology, 9,* 2. https://doi.org/10.3389/fmicb.2018.00002.

Morimoto, D., Tominaga, K., Nishimura, Y., Yoshida, N., Kimura, S., Sako, Y., et al. (2019). Cooccurrence of broad- and narrow-host-range viruses infecting the bloom-forming toxic cyanobacterium Microcystis aeruginosa. *Applied and Environmental Microbiology, 85,* e01170–19.

Morris, J. J., Kirkegaard, R., Szul, M. J., Johnson, Z. I., & Zinser, E. R. (2008). Facilitation of robust growth of Prochlorococcus colonies and dilute liquid cultures by "helper" heterotrophic bacteria. *Applied and Environmental Microbiology, 74,* 4530–4534. https://doi.org/10.1128/AEM.02479-07.

Morris, J. J., Lenski, R. E., & Zinser, E. R. (2012). The Black Queen Hypothesis: Evolution of dependencies through adaptive gene loss. *mBio, 3,* e00036–12. https://doi.org/10.1128/mBio.00036-12.

Needham, D. M., Chow, C. E. T., Cram, J. A., Sachdeva, R., Parada, A., & Fuhrman, J. A. (2013). Short-term observations of marine bacterial and viral communities: Patterns, connections and resilience. *The ISME Journal, 7,* 1274–1285. https://doi.org/10.1038/ismej.2013.19.

Needham, D. M., & Fuhrman, J. A. (2016). Pronounced daily succession of phytoplankton, archaea and bacteria following a spring bloom. *Nature Microbiology, 1,* 16005. https://doi.org/10.1038/nmicrobiol.2016.5.

Nelson, M. D., & Brand, E. L. (1979). Cell division periodicity in 13 species of marine phytoplankton on a light:dark cycle. *Journal of Phycology, 15,* 67–75.

Njus, D., Van Gooch, D., & Hastings, J. W. (1981). Precision of the Gonyaulax circadian clock. *Cell Biochemistry and Biophysics, 3,* 223–231. https://doi.org/10.1007/BF02782625.

Noordally, Z. B., & Millar, A. J. (2015). Clocks in algae. *Biochemistry, 54,* 171–183. https://doi.org/10.1021/bi501089x.

Oesterhelt, D., & Stoeckenius, W. (1973). Functions of a new photoreceptor membrane. *Proceedings of the National Academy of Sciences of the United States of America, 70,* 2853–2857. https://doi.org/10.1073/pnas.70.10.2853.

Olson, D. K., Yoshizawa, S., Boeuf, D., Iwasaki, W., & Delong, E. F. (2018). Proteorhodopsin variability and distribution in the North Pacific subtropical gyre. *The ISME Journal, 12,* 1047–1060. https://doi.org/10.1038/s41396-018-0074-4.

Ottesen, E. A., Young, C. R., Eppley, J. M., Ryan, J. P., Chavez, F. P., Scholin, C. A., et al. (2013). Pattern and synchrony of gene expression among sympatric marine microbial populations. *Proceedings of the National Academy of Sciences of the United States of America, 110,* E488–E497. https://doi.org/10.1073/pnas.1222099110.

Ottesen, E. A., Young, C. R., Gifford, S. M., Eppley, J. M., Marin, R., Schuster, S. C., et al. (2014). Multispecies diel transcriptional oscillations in open ocean heterotrophic bacterial assemblages. *Science, 345,* 207–212. https://doi.org/10.1126/science.1252476.

Pagarete, A., Chow, C. E. T., Johannessen, T., Fuhrman, J. A., Thingstad, T. F., & Sandaa, R. A. (2013). Strong seasonality and interannual recurrence in marine myovirus communities. *Applied and Environmental Microbiology, 79*, 6253–6259. https://doi.org/10.1128/AEM.01075-13.

Philosof, A., Battchikova, N., Aro, E. M., & Béjà, O. (2011). Marine cyanophages: Tinkering with the electron transport chain. *The ISME Journal, 5*, 1568–1570. https://doi.org/10.1038/ismej.2011.43.

Puxty, R. J., Millard, A. D., Evans, D. J., & Scanlan, D. J. (2015). Shedding new light on viral photosynthesis. *Photosynthesis Research, 126*, 71–97. https://doi.org/10.1007/s11120-014-0057-x.

Reppert, S. M., & Weaver, D. R. (2002). Coordination of circadian clocks in mammals. *Nature, 418*, 935–941.

Ribalet, F., Swalwell, J., Clayton, S., Jiménez, V., Sudek, S., Lin, Y., et al. (2015). Light-driven synchrony of *Prochlorococcus* growth and mortality in the subtropical Pacific gyre. *Proceedings of the National Academy of Sciences, 112*, 8008–8012. https://doi.org/10.1073/pnas.1424279112.

Ringelberg, J. (1995). Changes in light intensity and Diel vertical migration: A comparison of marine and freshwater environments. *Journal of the Marine Biological Association of the United Kingdom, 75*, 15–25. https://doi.org/10.1017/S0025315400015162.

Roenneberg, T., Nakamura, H., & Hastings, J. W. (1988). Creatine accelerates the circadian clock in a unicellular alga. *Nature, 334*, 432–434. https://doi.org/10.1038/334432a0.

Roenneberg, T., & Taylor, W. (1994). Light-induced phase responses in Gonyaulax are drastically altered by creatine. *Journal of Biological Rhythms, 9*, 1–12. https://doi.org/10.1177/074873049400900101.

Roux, S., Brum, J. R., Dutilh, B. E., Sunagawa, S., Duhaime, M. B., Loy, A., et al. (2016). Ecogenomics and potential biogeochemical impacts of globally abundant ocean viruses. *Nature, 537*, 689–693. https://doi.org/10.1038/nature19366.

Roy, S., Beauchemin, M., Dagenais-Bellefeuille, S., Letourneau, L., Cappadocia, M., & Morse, D. (2014). The Lingulodinium circadian system lacks rhythmic changes in transcript abundance. *BMC Biology, 12*, 1–8. https://doi.org/10.1186/s12915-014-0107-7

Seyedsayamdost, M. R., Case, R. J., Kolter, R., & Clardy, J. (2011). The Jekyll-and-Hyde chemistry of phaeobacter gallaeciensis. *Nature Chemistry, 3*, 331–335. https://doi.org/10.1038/nchem.1002.

Seymour, J. R., Amin, S. A., Raina, J. B., & Stocker, R. (2017). Zooming in on the phycosphere: The ecological interface for phytoplankton-bacteria relationships. *Nature Microbiology, 2*, 17065. https://doi.org/10.1038/nmicrobiol.2017.65.

Sieradzki, E. T., Ignacio-Espinoza, J. C., Needham, D. M., Fichot, E. B., & Fuhrman, J. A. (2019). Dynamic marine viral infections and major contribution to photosynthetic processes shown by spatiotemporal picoplankton metatranscriptomes. *Nature Communications, 10*, 1169. https://doi.org/10.1038/s41467-019-09106-z.

Lindh, M. V., Sjöstedt, J., Andersson, A. F., Baltar, F., Hugerth, L. W., Lundin, D., et al. (2015). Disentangling seasonal bacterioplankton population dynamics by high-frequency sampling. *Environmental Microbiology, 17*, 2459–2476. https://doi.org/10.1111/1462-2920.12720.

Sournia, A. (1975). Circadian periodicities in natural populations of marine phytoplankton. in M.B. In F. S. Russell & M. B. T.-A. Yonge (Eds.), *Advances in marine biology* (pp. 325–389). Academic Press. https://doi.org/10.1016/S0065-2881(08)60460-5.

Stearns, D. E., & Forward, R. B. (1984). Copepod photobehavior in a simulated natural light environment and its relation to nocturnal vertical migration. *Marine Biology, 82*, 91–100. https://doi.org/10.1007/BF00392767.

Steindler, L., Schwalbach, M. S., Smith, D. P., Chan, F., & Giovannoni, S. J. (2011). Energy starved candidatus pelagibacter ubique substitutes light-mediated ATP production for endogenous carbon respiration. *PLoS One, 6,* 1–10. https://doi.org/10.1371/journal.pone.0019725.

Straub, C., Quillardet, P., Vergalli, J., de Marsac, N. T., & Humbert, J. F. (2011). A day in the life of Microcystis aeruginosa strain PCC 7806 as revealed by a transcriptomic analysis. *PLoS One, 6,* e16208. https://doi.org/10.1371/journal.pone.0016208.

Strom, S. L. (2001). Light-aided digestion, grazing and growth in herbivorous protists. *Aquatic Microbial Ecology, 23,* 253–261. https://doi.org/10.3354/ame023253.

Sullivan, M. B., Huang, K. H., Ignacio-Espinoza, J. C., Berlin, A. M., Kelly, L., Weigele, P. R., et al. (2010). Genomic analysis of oceanic cyanobacterial myoviruses compared with T4-like myoviruses from diverse hosts and environments. *Environmental Microbiology, 12,* 3035–3056. https://doi.org/10.1111/j.1462-2920.2010.02280.x.

Sullivan, M. B., Lindell, D., Lee, J. A., Thompson, L. R., Bielawski, J. P., & Chisholm, S. W. (2006). Prevalence and evolution of core photosystem II genes in marine cyanobacterial viruses and their hosts. *PLoS Biology, 4,* 1344–1357. https://doi.org/10.1371/journal.pbio.0040234.

Suttle, C. A. (2005). Viruses in the sea. *Nature, 437,* 356–361. https://doi.org/10.1038/nature04160.

Suttle, C. A. (2007). Marine viruses—major players in the global ecosystem. *Nature Reviews. Microbiology, 5,* 801–812. https://doi.org/10.1038/nrmicro1750.

Swan, B. K., Tupper, B., Sczyrba, A., Lauro, F. M., Martinez-Garcia, M., Gonzàlez, J. M., et al. (2013). Prevalent genome streamlining and latitudinal divergence of planktonic bacteria in the surface ocean. *Proceedings of the National Academy of Sciences of the United States of America, 110,* 11463–11468. https://doi.org/10.1073/pnas.1304246110.

Tarangkoon, W., & Hansen, P. J. (2011). Prey selection, ingestion and growth responses of the common marine ciliate Mesodinium pulex in the light and in the dark. *Aquatic Microbial Ecology, 62,* 25–38. https://doi.org/10.3354/ame01455.

Teeling, H., Fuchs, B. M., Becher, D., Klockow, C., Gardebrecht, A., Bennke, C. M., et al. (2012). Substrate-controlled succession of marine bacterioplankton populations induced by a phytoplankton bloom. *Science (80-.), 336,* 608–611. https://doi.org/10.1126/science.1218344.

Teeling, H., Fuchs, B. M., Bennke, C. M., Krüger, K., Chafee, M., Kappelmann, L., et al. (2016). Recurring patterns in bacterioplankton dynamics during coastal spring algae blooms. *eLife, 5,* 1–31. https://doi.org/10.7554/eLife.11888.

Thompson, L. R., Zeng, Q., & Chisholm, S. W. (2016). Gene expression patterns during light and dark infection of Prochlorococcus by cyanophage. *PLoS One, 11,* e0165375. https://doi.org/10.1371/journal.pone.0165375.

Thompson, L. R., Zeng, Q., Kelly, L., Huang, K. H., Singer, A. U., Stubbe, J., et al. (2011). Phage auxiliary metabolic genes and the redirection of cyanobacterial host carbon metabolism. *Proceedings of the National Academy of Sciences of the United States of America, 108,* E757–E764. https://doi.org/10.1073/pnas.1102164108.

Tsai, A. Y., Gong, G. C., Sanders, R. W., Chiang, K. P., Huang, J. K., & Chan, Y. F. (2012). Viral lysis and nanoflagellate grazing as factors controlling diel variations of Synechococcus spp. summer abundance in coastal waters of Taiwan. *Aquatic Microbial Ecology, 66,* 159–167. https://doi.org/10.3354/ame01566.

Uzuka, A., Kobayashi, Y., Onuma, R., Hirooka, S., Kanesaki, Y., Yoshikawa, H., et al. (2019). Responses of unicellular predators to cope with the phototoxicity of photosynthetic prey. *Nature Communications, 10,* 1–17. https://doi.org/10.1038/s41467-019-13568-6.

Vaulot, D., Marie, D., Olson, R. J., & Chisholm, S. W. (1995). Growth of Prochlorococcus, a photosynthetic prokaryote, in the equatorial Pacific Ocean. *Science (80-.), 268,* 1480–1482. https://doi.org/10.1126/science.268.5216.1480.

Vislova, A., Sosa, O. A., Eppley, J. M., Romano, A. E., & DeLong, E. F. (2019). Diel oscillation of microbial gene transcripts declines with depth in Oligotrophic Ocean waters. *Frontiers in Microbiology, 10,* 2191. https://doi.org/10.3389/fmicb.2019.02191.

Vitaterna, M. H., Takahashi, J. S., & Turek, F. W. (2001). Overview of circadian rhythms. *Alcohol Research & Health, 25,* 85–93.

Waldbauer, J. R., Rodrigue, S., Coleman, M. L., & Chisholm, S. W. (2012). Transcriptome and proteome dynamics of a light-dark synchronized bacterial cell cycle. *PLoS One, 7,* e43432. https://doi.org/10.1371/journal.pone.0043432.

Welkie, D. G., Rubin, B. E., Diamond, S., Hood, R. D., Savage, D. F., & Golden, S. S. (2019). A hard day's night: Cyanobacteria in Diel cycles. *Trends in Microbiology, 27,* 231–242. https://doi.org/10.1016/j.tim.2018.11.002.

Whitman, W. B., Coleman, D. C., & Wiebe, W. J. (1998). Prokaryotes: The unseen majority. *Proceedings of the National Academy of Sciences of the United States of America, 95,* 6578–6583. https://doi.org/10.1073/pnas.95.12.6578.

Williamson, C. E., Fischer, J. M., Bollens, S. M., Overholt, E. P., & Breckenridgec, J. K. (2011). Toward a more comprehensive theory of zooplankton diel vertical migration: Integrating ultraviolet radiation and water transparency into the biotic paradigm. *Limnology and Oceanography, 56,* 1603–1623. https://doi.org/10.4319/lo.2011.56.5.1603.

Worden, A. Z., Follows, M. J., Giovannoni, S. J., Wilken, S., Zimmerman, A. E., & Keeling, P. J. (2015). Rethinking the marine carbon cycle: Factoring in the multifarious lifestyles of microbes. *Science, 347,* 1257594. https://doi.org/10.1126/science.1257594.

Yoshida, T., Nishimura, Y., Watai, H., Haruki, N., Morimoto, D., Kaneko, H., et al. (2018). Locality and diel cycling of viral production revealed by a 24 h time course cross-omics analysis in a coastal region of Japan. *The ISME Journal, 12,* 1287–1295. https://doi.org/10.1038/s41396-018-0052-x.

Young, M. W., & Kay, S. A. (2001). Time zones: A comparative genetics of circadian clocks. *Nature Reviews. Genetics, 2,* 702–715. https://doi.org/10.1038/35088576.

Yutin, N., & Koonin, E. V. (2012). Proteorhodopsin genes in giant viruses. *Biology Direct, 7,* 1 https://doi.org/10.1186/1745-6150-7-34.

Zhao, Z., Gonsior, M., Schmitt-Kopplin, P., Zhan, Y., Zhang, R., Jiao, N., et al. (2019). Microbial transformation of virus-induced dissolved organic matter from picocyanobacteria: Coupling of bacterial diversity and DOM chemodiversity. *The ISME Journal, 13,* 2551–2565. https://doi.org/10.1038/s41396-019-0449-1.

Zimmerman, A. E., Howard-Varona, C., Needham, D. M., John, S. G., Worden, A. Z., Sullivan, M. B., et al. (2019). Metabolic and biogeochemical consequences of viral infection in aquatic ecosystems. *Nature Reviews. Microbiology, 18,* 21–34. https://doi.org/10.1038/s41579-019-0270-x.

Zinser, E. R., Lindell, D., Johnson, Z. I., Futschik, M. E., Steglich, C., Coleman, M. L., et al. (2009). Choreography of the transcriptome, photophysiology, and cell cycle of a minimal photoautotroph, Prochlorococcus. *PLoS One, 4,* e5135. https://doi.org/10.1371/journal.pone.0005135.

CPI Antony Rowe
Chippenham, UK
2020-10-05 17:49